T0183478

Lecture Notes in Computer Science　　10494

Commenced Publication in 1973
Founding and Former Series Editors:
Gerhard Goos, Juris Hartmanis, and Jan van Leeuwen

Edmon Begoli · Fusheng Wang
Gang Luo (Eds.)

Data Management and Analytics for Medicine and Healthcare

Third International Workshop, DMAH 2017
Held at VLDB 2017
Munich, Germany, September 1, 2017
Proceedings

Editors
Edmon Begoli (ID)
Oak Ridge National Laboratory
Oak Ridge, TN
USA

Gang Luo
University of Washington
Seattle, WA
USA

Fusheng Wang
Stony Brook University
Stony Brook, NY
USA

ISSN 0302-9743   ISSN 1611-3349   (electronic)
Lecture Notes in Computer Science
ISBN 978-3-319-67185-7   ISBN 978-3-319-67186-4   (eBook)
DOI 10.1007/978-3-319-67186-4

Library of Congress Control Number: 2017952570

LNCS Sublibrary: SL3 – Information Systems and Applications, incl. Internet/Web, and HCI

Printed on acid-free paper

This Springer imprint is published by Springer Nature
The registered company is Springer International Publishing AG
The registered company address is: Gewerbestrasse 11, 6330 Cham, Switzerland

Preface

In this volume we present the accepted contributions for the Third International Workshop on Data Management and Analytics for Medicine and Healthcare (DMAH 2017), held in Munich, Germany, in conjunction with the 43rd International Conference on Very Large Data Bases (VLDB), on September 1st, 2017.

The goal of the workshop was to bring together researchers from the cross-cutting domains of research including information management and biomedical informatics. The workshop aimed to foster the exchange of information and discussions on innovative data management and analytics technologies. We encouraged topics that highlighted the end-to-end applications, systems and methods addressing problems in healthcare, public health, and everyday wellness; integration with clinical, physiological, imaging, behavioral, environmental, and "omics" data, as well as the data from social media and the Web. Our hope for this workshop was to provide a unique opportunity for mutual benefits and informative interaction between information management and biomedical researchers from the interdisciplinary fields.

A total of 16 papers were submitted to the DMAH workshop. A rigorous, single-blind peer-review selection mechanism was adapted, resulting in 9 accepted papers presented at the workshop. Each paper was reviewed by three members of the Program Committee, who were carefully selected for their knowledge and competence. As far as possible, papers were matched with the reviewers' particular interests and special expertise. The result of this careful process can be seen here in the high quality of the contributions published within this volume.

We would like to express our sincere thanks especially to the internationally renowned speakers who gave keynote talks at the workshop plenary sessions: Li Xiong from Emory University, USA, and Vagelis Hristidis from University of California, Riverside, USA.

We would like to thank the members of the Program Committee for their attentiveness, perseverance, and willingness to provide high-quality reviews.

July 2017

Edmon Begoli
Fusheng Wang
Gang Luo

Organization

Program Committee

Jesús B. Alonso-Hernández	Universidad de Las Palmas de Gran Canaria, Spain
Edmon Begoli	Oak Ridge National Laboratory, USA
Thomas Brettin	Argonne National Laboratory, USA
J. Blair Christian	Oak Ridge National Laboratory, USA
Carlo Combi	Università degli Studi di Verona, Italy
Kerstin Denecke	Bern University of Applied Sciences, Switzerland
Dejing Dou	University of Oregon, USA
Alevtina Dubovitskaya	EPFL, HES-SO, Switzerland
Peter Elkin	University at Buffalo, USA
Vijay Gadapally	MIT Lincoln Labs/CSAIL, USA
Zhe He	Florida State University, USA
Guoqian Jiang	Mayo Clinic College of Medicine, USA
Jun Kong	Emory University, USA
Tahsin Kurc	Stony Brook University, USA
Ulf Leser	Humboldt-Universität zu Berlin, Germany
Yanhui Liang	Stony Brook University, USA
Gang Luo	University of Washington, USA
Fernando Martin-Sanchez	Weill Cornell Medicine, USA
Wolfgang Mueller	Heidelberg Institute for Theoretical Studies, Germany
Casey Overby	Johns Hopkins University, USA
Fusheng Wang	Stony Brook University, USA
Hua Xu	The University of Texas School of Biomedical Informatics at Houston, USA

Abstracts of the Keynotes

Health Data Management and Analytics with Privacy and Confidentiality

Li Xiong

Emory University, Atlanta GA 30322, USA
lxiong@emory.edu

Abstract. Managing and analyzing large-scale clinical and public health data while protecting privacy of human subjects has been a key challenge in biomedical research. Traditional de-identification approaches are subject to various re-identification and disclosure risks and do not provide sufficient privacy protection for patients. This talk gives an overview of our work on privacy preserving health data sharing and analytics along two dimensions: (1) data encryption techniques that support secure computation and query processing on the encrypted data without disclosing the raw data, (2) differential privacy techniques that ensure the computation and query results do not disclose patient information. Focusing on the second dimension, a set of differential privacy techniques are presented that handle different types of data including relational, sequential, and time series data. Case studies using real health datasets are presented to demonstrate the feasibility of the solutions while outlining their limitations and open challenges.

Keywords: Health data management · Data analytics · Differential privacy · Data encryption · Secure computation

Acknowledgement. The presented work was supported by the National Institute of Health (NIH) under award number R01GM114612, R01GM118609-01A1, the Patient-Centered Outcomes Research Institute (PCORI) under contract ME-1310-07058.

Analysis of Online Health-Related User-Generated Content

Vagelis Hristidis

University of California, Riverside
vagelis@cs.ucr.edu

Abstract. An increasing amount of health-related content is posted online by patients, ranging from health forums to provider reviews. Analyzing this mostly text information can discover health trends and help patients make more informed decisions. We will discuss about the technical challenges involved in analyzing such data, including the use of biomedical ontologies, concept extraction, training set expansion and summarization. Completed and ongoing work on various application will be presented, as well as opportunities for future research.

Contents

Data Privacy and Trustability for Electronic Health Records

How Blockchain Could Empower eHealth: An Application for Radiation
Oncology (Extended Abstract)   3
    Alevtina Dubovitskaya, Zhigang Xu, Samuel Ryu, Michael Schumacher,
    and Fusheng Wang

Biomedical Data Management and Integration

On-Demand Service-Based Big Data Integration: Optimized
for Research Collaboration   9
    Pradeeban Kathiravelu, Yiru Chen, Ashish Sharma,
    Helena Galhardas, Peter Van Roy, and Luís Veiga

CHIPS – A Service for Collecting, Organizing, Processing,
and Sharing Medical Image Data in the Cloud   29
    Rudolph Pienaar, Ata Turk, Jorge Bernal-Rusiel, Nicolas Rannou,
    Daniel Haehn, P. Ellen Grant, and Orran Krieger

High Performance Merging of Massive Data from Genome-Wide
Association Studies   36
    Xiaobo Sun, Fusheng Wang, and Zhaohui Qin

An Emerging Role for Polystores in Precision Medicine   41
    Edmon Begoli, J. Blair Christian, Vijay Gadepally,
    and Stavros Papadopoulos

Online Mining of Health Related Data

Social Media Mining to Understand Public Mental Health   55
    Andrew Toulis and Lukasz Golab

Clinical Data Analytics

Effects of Varying Sampling Frequency on the Analysis of Continuous
ECG Data Streams   73
    Ruhi Mahajan, Rishikesan Kamaleswaran, and Oguz Akbilgic

Detection and Visualization of Variants in Typical Medical
Treatment Sequences   88
   Yuichi Honda, Muneo Kushima, Tomoyoshi Yamazaki,
   Kenji Araki, and Haruo Yokota

Umedicine: A System for Clinical Practice Support and Data Analysis   102
   Nuno F. Lages, Bernardo Caetano, Manuel J. Fonseca,
   João D. Pereira, Helena Galhardas, and Rui Farinha

Association Rule Learning and Frequent Sequence Mining of Cancer
Diagnoses in New York State   121
   Yu Wang and Fusheng Wang

Healthsurance – Mobile App for Standardized Electronic Health
Records Database   136
   Prateek Jain, Sagar Bhargava, Naman Jain, Shelly Sachdeva,
   Shivani Batra, and Subhash Bhalla

Author Index   155

Data Privacy and Trustability
for Electronic Health Records

How Blockchain Could Empower eHealth: An Application for Radiation Oncology

(Extended Abstract)

Alevtina Dubovitskaya[1,2(✉)], Zhigang Xu[3], Samuel Ryu[3],
Michael Schumacher[1], and Fusheng Wang[4]

[1] University of Applied Sciences Western Switzerland (HES-SO),
Sierre, VS, Switzerland
alevtina.dubovitskaya@hevs.ch
[2] École Polytechnique Fédérale de Lausanne (EPFL), Lausanne, VD, Switzerland
[3] Stony Brook Medicine, Stony Brook, NY, USA
[4] Stony Brook University (SBU), Stony Brook, NY, USA

Abstract. Electronic medical records (EMRs) contain critical, highly sensitive private healthcare information, and need to be frequently shared among peers. Blockchain provides a shared, immutable and transparent history of all the transactions to build applications with trust, accountability and transparency. This provides a unique opportunity to develop a secure and trustable EMR data management and sharing system using blockchain. In this paper, we discuss our perspectives on blockchain based healthcare data management and present a prototype of a framework for managing and sharing EMR data for cancer patient care.

Keywords: eHealth · EMR · Blockchain · Radiation oncology

1 Introduction

Patient's healthcare data are often distributed among different actors of the Healthcare system. Patients need to share the data during the treatment, or for the purposes of medical research. Medical data are highly sensitive and when sharing or transferring them from one institution to another (for primary care and for research purposes), according to the legislation in Europe and USA a patient has to provide a consent where the access control policy is defined. Consents are usually not personalized standard forms, and it is not easy for the patient to express his access control policy. For example, when a patient wishes to receive an independent opinion about his condition from another medical doctor in the same medical institution, or prefers not to share some part of his medical history.

In case of the data aggregation for the research purposes, patients could be concerned about violation of their privacy and may not provide a consent to share their data as they are. An alternative to the consent collection is data

E. Begoli et al. (Eds.): DMAH 2017, LNCS 10494, pp. 3–6, 2017.
DOI: 10.1007/978-3-319-67186-4_1

anonymization. However, the privacy violation could happen in case of linking multiple datasets containing anonymized information about the same patient [6]. How to keep track of the shared information about the patient in a longterm perspective?

A patient with chronic disease or serious medical condition may need to keep track of his patient's record through his life, or may need to delegate the data management to his relatives due to the medical condition. Management of medical history, access control, prescriptions, medical expenses, insurance correspondence and payments is unavoidable notoriously time-consuming and bothersome for most of such patients. For example, treatment of a cancer patient may urgently require knowing the radiation dose received during the life-long treatment. Consent management and data transfer may delay a critical treatment. The goal of this work is to develop a framework that can be used by patients, medical doctors and other entities involved in healthcare processes for patient's data management. We present a solution that tackles the problems discussed above and ensures privacy, security, availability, and fine-grained access control over EMR data.

2   Potential Applications of Blockchain in eHealth

Background knowledge and related work. Blockchain – is a distributed ledger technology based on the principles of peer-to-peer network and cryptographic primitives (such as hash, asymmetric encryption and digital signature) [1]. Having access to a ledger - shared, immutable, and transparent history of all the actions (transactions) performed by participants of the network (such as a patient modifying permissions, a doctor, accessing or uploading new data, or sharing them for research) overcome the issues presented above. Based on how the identity of a user is defined within a network, one could distinguish between permissioned and permissionless blockchain systems. The potential of the applications built on top of the blockchain technology for healthcare data management has been recently discussed by researchers. MedRec is the first and the only functioning prototype that have been proposed recently [2]. Ariel et al. presented a system based on permissionless blockchain implementation. Our prototype presented further in this manuscript significantly differs from the framework in [2]. First, we have chosen to use a permissioned blockchian technology that provides better protection of the patient's privacy, do not involves transaction fees and "mining". Second, in [2] the patients data are stored locally at every node, which does not ensure availability of the data in case if the hospital node is temporary off-line.

Applications in eHealth. Blockchain provides a unique opportunity to support healthcare. Hereafter we discuss a blockchain application to support Connected health [4]. Sharing the ledger (using the permission-based approach) among entities (such as medical doctors, medical institutions, insurance companies and pharmacies) will facilitate medication and cost management for a patient, especially in case of chronic disease management. Providing pharmacies with accurately updated data about prescriptions will improve the logistics. Access to a common

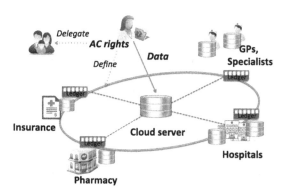

Fig. 1. Connecting different healthcare players for better patient care.

ledger would allow the transparency in the whole process of the treatment, from monitoring if a patient follows correctly the prescribed treatment, to facilitating communication with an insurance company regarding the costs of the treatment and medications. The network would be formed by the trusted peers. The peers will run consensus protocol and maintain a distributed ledger. The patient (or his relatives) will be able to access and manage the data through a web interface (Fig. 1). The key management and the access control policy will be encoded in a chaincode, thus, ensuring data security and patients privacy.

In case of medical data aggregation for research purposes it is highly important to ensure that the sources of the data are trusted medical institutions and, therefore, the data are authentic. Using shared distributed ledger will provide tracebility and will guarantee patient's privacy as well as the transparency of the data aggregation process. Due to the current lack of appropriate mechanisms, patients are often unwilling to participate in data sharing. Using blockchain technology will facilitate the process of collecting patients data for research purposes.

3   Application in Radiation Oncology

We apply the blockchain technology to create a prototype of an oncology-specific clinical data sharing system. To present our solution, we take as an example an oncology information system, ARIA [5], which is widely used to facilitate oncology-specific comprehensive information and images management. ARIA combines radiation, medical and surgical oncology information and can assist clinicians to manage different kinds of medical data, develop oncology-specific care plans, and monitor radiation dose received by patients.

Proposed framework. Hereafter we describe an architecture of the framework for radiation oncology data management. To develop a prototype we used Hyperledger Fabric – open source implementation of the permissioned blockchain technology [3]. Our architecture consists of a user interface and a backend that is composed of the following components: membership service and certification authority, network of nodes (deployed in the medical institutions and connected to the database), load balancer to redirect a user to any of the trusted nodes in the network, separate cloud-based storages for patient's data and certificates.

Functionality of the prototype. The functionality of the prototype is the following. A patient can register in the system (via membership service), generate a secret key (AES), public/private key pair, receive a certificate from the certification authority. Then the patient can login and create his record. In order to define access control policies he submits a transaction that will specify which doctor is able to access which type of the data within specified time interval. The patient could also upload the data to the cloud repository: after encrypting them with his own secret key and hashing to ensure the data integrity. The metadata will be stored on the blockchain: the transaction will contain the hash of the file, a URL of the file, id of the patient that uploaded the file. To provide an access to the data the patient's key has to be shared using the certificate of the doctor. A doctor also needs to register in the system, generate a public/private key pair, and obtain a certificate from the certification authority. Similarly to the patients medical doctors are able to upload the data about the patient, and access them but only based on the permissions specified by the patients.

4   Conclusion and Future Work

We proposed potential applications of blockchain technology in healthcare data management. Based on the requirements from medical perspective we implemented a prototype of a framework for data management and sharing in the oncology patient care. We are going to test our framework with the patients data in hospital environment. Extending metadata representations, adding new categories (e.g., images), evaluation and improvement of the user interfaces, introducing new actors (researchers, pharmacologists, insurances) are the next steps of our work. Using blockchain technology allows to ensure privacy, security, availability, and fine-grained access control over EMR data. The proposed work can significantly reduce the turnaround time for EMR sharing, improve decision making for medical care, and reduce the overall cost.

References

1. Nakamoto, S.: Bitcoin: a Peer-to-Peer electronic cash system (2008). https://bitcoin.org/bitcoin.pdf
2. Azaria, A., Ekblaw, A., Vieira, T., Medrec, L.A.: Using blockchain for medical data access and permission management. In: 2016 2nd International Conference on Open and Big Data (OBD), pp. 25–30, August 2016
3. Cachin, C.: Architecture of the hyperledger blockchain fabri (2016)
4. Connected health. https://en.wikipedia.org/
5. ARIA Oncology Information System. Varian Medical Systems. https://www.varian.com/oncology/products/software/information-systems/aria-ois-radiation-oncology. Accessed 9 Mar 2017
6. Baig, M.M., Li, J., Liu, J., Wang, H.: Cloning for privacy protection in multiple independent data publications. In: Proceedings of the 20th ACM International Conference on Information and Knowledge Management - CIKM 2011, p. 885 (2011)

Biomedical Data Management
and Integration

On-Demand Service-Based Big Data Integration: Optimized for Research Collaboration

Pradeeban Kathiravelu[1,2](✉), Yiru Chen[3], Ashish Sharma[4],
Helena Galhardas[1], Peter Van Roy[2], and Luís Veiga[1]

[1] INESC-ID/Instituto Superior Técnico, Universidade de Lisboa, Lisbon, Portugal
{pradeeban.kathiravelu,helena.galhardas,luis.veiga}@tecnico.ulisboa.pt
[2] Université catholique de Louvain, Louvain-la-Neuve, Belgium
peter.vanroy@uclouvain.be
[3] Peking University, Beijing, China
chen1ru@pku.edu.cn
[4] Emory University, Atlanta, GA, USA
ashish.sharma@emory.edu

Abstract. Biomedical research requires distributed access, analysis, and sharing of data from various disperse sources in the Internet scale. Due to the volume and variety of big data, materialized data integration is often infeasible or too expensive including the costs of bandwidth, storage, maintenance, and management. *Óbidos* (**O**n-demand **B**ig **D**ata **I**ntegration, **D**istribution, and **O**rchestration **S**ystem) provides a novel on-demand integration approach for heterogeneous distributed data. Instead of integrating data from the data sources to build a complete data warehouse as the initial step, *Óbidos* employs a hybrid approach of virtual and materialized data integrations. By allocating unique identifiers as pointers to virtually integrated data sets, *Óbidos* supports efficient data sharing among data consumers. We design *Óbidos* as a generic service-based data integration system, and implement and evaluate a prototype for multimodal medical data.

Keywords: Materialized data integration · Virtual data integration · Integrated data repository · Extract, Transform, Load (ETL) · Big data integration · DICOM

1 Introduction

Scale and diversity of big data is on rise, with geographically distributed data of exabytes. Integration of data is crucial for various application domains ranging from medical research [1] to transport planning [2], to enable data analysis, finding relationships, and retrieving information, across data from various sources. However, storing the integrated data in a separate data repository can be deferred to query time or avoided altogether. This approach minimizes outdated data in the integrated data repository, and mitigates the maintenance costs associated with frequent data refreshments and data cleaning processes.

© Springer International Publishing AG 2017
E. Begoli et al. (Eds.): DMAH 2017, LNCS 10494, pp. 9–28, 2017.
DOI: 10.1007/978-3-319-67186-4_2

Disparate multimodal medical data [3] comprising imaging, clinical, and genomic data is published by data producers such as physicians and health personnel. Data of each of these domains (including clinical, pathology, and other subdomains of medical data research) are of heterogeneous types and various formats - textual, binary, or a hybrid of both. Medical data science researchers need to consume data sets spanning multiple data sources. This typically requires integration and analysis of multimodal medical data.

Adhering to standard formats such as DICOM (The Digital Imaging and Communications in Medicine) [4] and NIfTI (Neuroimaging Informatics Technology Initiative) [5], each medical image consists of binary raw image data and textual metadata. This textual metadata, present in many file formats that support a large-scale binary data, permit a faster analysis or provide a brief overview of the respective binary data. Most of the medical research studies access or query only the metadata at least till a certain subset is identified for further detailed analysis. Moreover, each medical data research is typically limited to certain sets of data from the data sources. Therefore, the researchers do not need to integrate entirely the data sources, which is also infeasible considering the volume of the data. While the data sets of interest vary between the researchers, the subsets of data also need to be shared among researchers for collaborations and to repeat the experiments for verifiability of scientific and medical research.

Traditional/eager ETL (Extract, Transform, Load) process constructs the entire data warehouse as the initial process. This bootstrapping process comprising the initial loading is lengthy and the study-specific researches are unnecessarily delayed when only a small subset of data is relevant to the queries. Lazy ETL [6] avoids the lengthy bootstrapping process by eagerly loading only the metadata while the actual data is loaded in a lazy manner. The actual data is accessed only at query time. Hence, bootstrapping in lazy ETL is very fast. Nevertheless, lazy ETL approach does not offer scalability to store, access, and share big data efficiently across multiple data consumers.

Motivation: Consider a research to study the effects of a medicine to treat brain tumor in patients of certain age groups against that of a placebo. This research involves a large amount of data consisting of cancer images, clinical records, lab measurements, and observations, collected over past several months. Though the data is huge, the metadata often gives an overall description of the binary data, sufficient for at least the early stages of a research study. When performing data analysis, loading of binary data can be deferred, analysing the textual metadata first to find the required binary data. Hence, loading of cancer images for integration and analysis should be deferred to a latter phase, loading the relevant sets of metadata at first. This approach is more computation and bandwidth efficient, compared to loading all the binary data at once.

The data sets of interest are well-defined by the medical researcher as a small sub set of a much larger data from multiple data sources. The existing data integration approaches are unfit for the study-specific medical research, as these approaches do not leverage the domain knowledge of the researcher, in filtering the data early on. On the other hand, manually searching, download-ing, and integrating data is repetitive, time consuming, and expensive. Hence, medical research requires an approach to (i) incrementally construct a study-specific integrated data repository on-demand, (ii) share interesting subsets of data with minimal bandwidth consumption, (iii) track the loading of data from data sources to load only the changes or updates, and (iv) minimize near dupli-cates due to corrupted or moved data in the integrated data repository, or changes/updates in the data sources.

Contribution: We present *Óbidos*[1], an on-demand data integration system for heterogeneous big data, and implement its prototype for medical research, addressing the identified shortcomings. *Óbidos* follows a hybrid approach in load-ing and accessing data. Following lazy ETL, it avoids eager loading of binary data, exploiting the metadata in accessing and integrating binary data from data sources.

Óbidos leverages the well-defined hierarchical structure of medical data of formats such as DICOM to construct subsets of data that are of interest. *Óbidos* loads only a subset of metadata, unlike lazy ETL that advocates an eager loading of metadata. First, *Óbidos* finds a superset of data, including that of the data sets of interest. Then it loads the metadata of the chosen superset, for example: (i) in a hierarchical data structure such as DICOM, the chosen subset consists of an entire level of granularity that has the data that is of interest to the tenant; (ii) in a file system, it may include contents of an entire directory. By leveraging the data loading APIs offered by the respective data source platforms, only the chosen set of metadata is queried, instead of querying the whole data sources, which is not practical and redundant.

The integrated data access tends to be slower in Lazy ETL as it only loads metadata till the query is processed, whereas the associated binary data is much larger in volume and will consume a large time to load from the data sources depending on the network bandwidth. *Óbidos* overcomes this by having a mate-rialized data integration for the data that is the outcome of the queries, in a scalable storage with an efficient approach to access the data. Furthermore, *Óbidos* leverages the metadata in tracking the changes and downloads from the data sources.

Óbidos *Orchestration:* For loading and sharing of medical research data, we envision *Óbidos* to be deployed in multiple research institutes. Each such insti-tute is called a data consumer, who needs to integrate and consume research data from multiple data sources, and share among one another. *Óbidos* is to be typically hosted in a cluster of computers in a research institute as a single yet, distributed deployment. Each such deployment is defined as a data consumer deployment, that is used by multiple researchers, who are called the tenants of the *Óbidos* deployment.

---

[1] *Óbidos* is a medieval fortified town that has been patronized by various Portuguese queens. It is known for its sweet wine, served in a chocolate cup.

Óbidos identifies data that is integrated by the respective tenant. Through the service-based interface of *Óbidos*, tenants can access the data in their respective *Óbidos* deployment, or data integrated in a remote *Óbidos* deployment. Customizable to the policies of the research institutes and the needs of the users, *Óbidos* orchestrates the way research data is shared among the users. In addition to colleagues in a research institute sharing data among them, peers from other institutes may invoke the web service interface of *Óbidos* to access the shared data in any *Óbidos* deployment. The service-based approach facilitates protected access through the Internet to the integrated data repository of *Óbidos* data consumer deployment.

The rest of this paper is structured as follows: Sect. 2 discusses the background behind the multimodal medical data integration and *Óbidos* approach. Section 3 elaborates the deployment and solution architecture of *Óbidos*. Section 4 discusses how *Óbidos* prototype is built for multimodal medical research data. Section 5 evaluates *Óbidos* for its performance. Section 6 discusses the related work on service-based and virtual data integration platforms. Finally, Sect. 7 provides a summary of the current state and future research work based on *Óbidos*.

2   Background

In this section, we look at the motivations behind *Óbidos* and its contributions, with a background in medical data access, integration, and sharing.

Data integration is performed traditionally by ingesting data from various sources into a data warehouse and processing it afterwards. Due to the proliferation of binary data, the associated textual metadata often replaces the data itself in indexing, identifying, and loading the relevant data. In multidisciplinary research, the number of potentially relevant data sources are monotonically increasing. Therefore, a massive amount of data or metadata need to be loaded initially, if a data warehouse is to be constructed from the entire data sources as the first step. This leads to a waste of time and storage where a large fraction of data that is useless and irrelevant is loaded in the beginning. This can be avoided by improving the lazy loading approach of loading only the metadata with the research-specific knowledge on the data sets of interest. Instead of loading all the metadata from the sources, we advocate loading only the metadata corresponding to the datasets of interest, or a superset of them (rather than loading the entire metadata, which is infeasible). Users may specify all the data sets of interest to load the relevant metadata. Alternatively, it should also be possible to infer their interest from the search queries.

Óbidos attempts to solve two major challenges in research big data integration and sharing: (i) the challenges of volume and variety associated with the distributed landscape of medical research data and the data sources, and (ii) the need for a scalable implementation to integrate and share the big data between various researchers.

2.1   Integrating Distributed and Disperse Data

Big data integration needs to consider tens of thousands of data sources even in a single domain [7]. Multi-domain and cross-disciplinary researches have left us with a much larger and diverse number of data sources to be accessed and integrated, compared to the traditional data integration requirements. Velocity, variety, and veracity aspects of big data present further challenges in addition to the volume of the data involved. Hence, if possible, a subset of the data needs to be chosen in advance for the integration, instead of a complete offline warehousing from all the available data sources [7].

The very nature of dispersed data from various data repositories poses a set of issues that need to be addressed. Due to the increasing number and proliferation of data repositories, it is not only unnecessary, but also infeasible to access all the data or even the metadata. The repositories offer protected access to the data they host, typically through a web interface such as web services or web applications. They present a hierarchical or indexed view of the data they host, despite the data itself being often binary and unstructured. These respective interfaces of data repositories need to be leveraged in efficiently accessing and loading data. Traditional approach of loading an entire data or metadata is not applicable in service-based data platforms.

2.2   Scalable Implementation for Big Data Processing

With the increasing scale and complexity of data storage and analysis, it becomes evident that each task of an ETL process should be executed in a parallel manner appropriately, for a faster completion [8]. Distributing the ETL process may reduce latency in accessing, transforming, and loading data into a data warehouse [9], thus preventing main memory from becoming a bottleneck in data transformations at any point. In-Memory Data Grid (IMDG) platforms [10] can be leveraged in data integration and data quality platforms to improve performance and scalability. For example, $\partial u \partial u$ [11] provides a multi-tenant in-memory distributed execution platform for near duplicate detection algorithms for data integration from multiple sources.

The data sources consist of heterogeneous data, often unstructured or semi-structured. Hence, the integrated data should be stored in a scalable storage that supports storing of unstructured data. Thus, typical medical data integration solutions that restrict themselves to well-defined schemas are not suitable for integrating medical research big data. As NoSQL solutions allow storing unstructured data, they should be leveraged along with an architecture that supports indexing of the stored binary data. Hence, even originally unrelated - structured and unstructured data (for example, pathology images and tables of patient records) can be loaded together for a specific research.

The integrated data should be offered with a secured access due to the sensitive nature of medical data. API gateways such as Tyk[2], Kong[3], and API

---

[2] https://tyk.io/.
[3] https://getkong.org/.

Umbrella[4] offer protected access to the APIs, with authentication and authorization to the APIs that access the underlying data. By offering a service-based data access and integration, data can be protected from direct access, by leveraging such API gateways, in contrast to offering a direct access to the integrated data repository.

3   *Óbidos* Data Integration System

Óbidos offers an on-demand data integration process for data in a hierarchical structure. It comprises of a scalable storage: (i) quicker access to metadata stored in-memory, and (ii) much larger distributed physical storage for the materialized binary data. *Óbidos* comprises a set of workflows that are initiated by invoking its APIs.

3.1   *Óbidos* Distributed Storage and Deployment

Óbidos offers a hybrid of materialized and virtual data integration. It stores previously accessed and loaded data of interest in an integrated data repository in Apache Hadoop Distributed File System (HDFS) [12] for subsequent accesses. HDFS is used to store the integrated data, due to its scalability and support for storing unstructured and semi-structured, binary and textual data. It eagerly loads only the metadata for the objects of the chosen subsets of data at a specified granularity, into an In-Memory Data Grid of Infinispan [13]. For example, DICOM is structured from collections, patients, studies, series, to images - from coarse to fine granularity. Hence, if a specific study is chosen, all the metadata associated to the study will be loaded, indicating that all the series and images belonging to the chosen study are of interest to the tenant. The integrated data is shared through public identifiers, among tenants inside or between data consumers, via a web services interface. Furthermore, *Óbidos* offers an efficient and fast access to the integrated and virtually integrated data repository in SQL queries.

The metadata of the binary data in HDFS is stored in tables hosted in Apache Hive [14] metastore based on HDFS. Hive tables are indexed for efficient user queries to identify and locate the binary data in the HDFS integrated data repository which is typically unstructured. We call this metadata that operates as an index to the unstructured storage of HDFS, **"metametadata"** to differentiate it from the metadata of larger scope that is stored in the virtual data integration layer (as this indexes data and metadata loaded from the sources). Various storages are unified and accessed seamlessly by leveraging Apache Drill [15], as it enables SQL queries on structured, semi-structured, and unstructured data.

Figure 1 illustrates the deployment landscape of data consumers and data sources. Each data consumer consists of a distinct *Óbidos* deployment. Each of the *Óbidos* deployment has multiple tenants accessing, loading, and integrating

---

4 https://apiumbrella.io/.

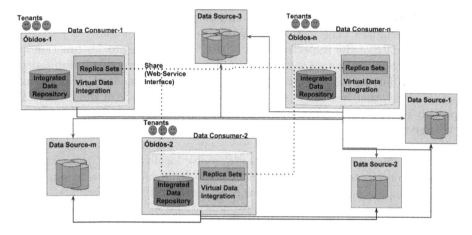

Fig. 1. Landscape of *Óbidos* data consumer deployments, along with the data sources

data. Tenants of *Óbidos* data consumers access data from the data sources to build their study-specific integrated data repository. When a tenant searches or queries the data sources, a Universal Unique Identifier (UUID) is created to point to the subsets of data chosen/accessed by the tenant, spanning the multiple data sources. These pointers are generated by appending a locally created unique identifier to the public IP address or Uniform Resource Identifier (URI) of the *Óbidos* deployment server. We call these pointers **"replica sets"**. Replica sets can be shared by the tenant that who created them among the other tenants of the same deployment, or to other *Óbidos* data consumer deployments. Hence, the relevant research data can be shared without copying the actual binary data or metadata.

Data sharing in *Óbidos* works like the Dropbox shared links; no actual data is copied and shared. Only a link that points to the data in data sources, is shared - either directly through the *Óbidos* web service API invocations or outside *Óbidos* (for example, sharing the replica set through other communication media such as email). Once a replica set shared by the tenant is accessed by a remote tenant, the replica set will be resolved into pointers to the actual data sources. The relevant data pointed by the shared replica set can later be loaded by the remote tenants to their deployment. In addition to one-to-one sharing of data between two tenants of the same or different *Óbidos* deployments (deployed in two research institutes), a tenant may opt to create a few replica sets with public access. These public replica sets can be searched and accessed through a public web service API, as they are indexed and stored accordingly.

Tenants can access the data residing in the integrated data repository of their data consumer deployment. Hence, if a replica set is shared among the tenants in the same deployment, it is resolved locally. If the data exists in the integrated data repository, the tenants may access it directly; otherwise, the data is loaded for future accesses.

Óbidos *Virtual Proxies:* The virtual data integration space consists of virtual proxies of metadata. The virtual proxy of an object is essentially a partially loaded metadata of the respective data object. The virtual proxies function as a value holder for the actual metadata, and have sufficient minimal information on metadata for the search query. If only a fraction of metadata is relevant for a search query, that fraction is sufficient as the virtual proxy of the metadata, especially when the metadata itself is of a complex or hierarchical data structure or is large in volume.

When a virtual proxy is accessed by a query, the respective metadata is loaded from the data sources, if it does not resolve locally (means, the complete metadata object is not present yet in the Óbidos deployment). This incremental replication approach is motivated by our previous work [16] on achieving efficient replication of larger objects in a wide area network (WAN). This enables Óbidos efficient loading of complex metadata faster than the lazy ETL approaches.

Near Duplicate Detection: As the data comes from multiple sources there are possibilities for duplicates. A distributed near duplicate detection algorithm is executed on the metadata periodically, or on-demand, to detect the matches from the heterogeneous sources in building up the integrated data repository. The detected near duplicate pairs are stored in a separate data source, which may be leveraged to improve the accuracy of the detected duplicates. This further allows fixing false positives of detected duplicate pairs in a latter iteration. Óbidos compares the timestamps from the metadata of the relevant subsets of data in the data sources against that of the loaded data. Only the updates are loaded when a data set has already been loaded to the Óbidos integrated data repository. While previously loaded binary data can be compared with the corresponding data from the source to identify or confirm corruption, typically, the textual metadata is used for the duplicate detection. We exploit our previous work [11] to distribute classic near duplicate detection algorithms such as PPJoin and PPJoin+ [17] in a cluster, to execute and detect near duplicates among the metadata. Thus, we identify duplicates among the respective loaded and source data.

3.2 *Óbidos* Software Architecture

Óbidos is used by (i) the tenants (the regular users), (ii) administrators that deploy and configure Óbidos, and (iii) tenants of other Óbidos deployments who have a limited web service-based access. Figure 2 depicts a high-level solution architecture of Óbidos, representing any Óbidos data consumer deployment depicted in Fig. 1. Through its northbound, eastbound, southbound, and westbound APIs, Óbidos offers capabilities of data access and integration, configuration and management, storage integration, and data sources federation, respectively.

The **Virtual Data Integration Layer** stores the virtually integrated metadata in-memory in an Infinispan data grid. Óbidos **Integrated Data Repository** consists of the actual storage behind the **Materialized Data Integration**

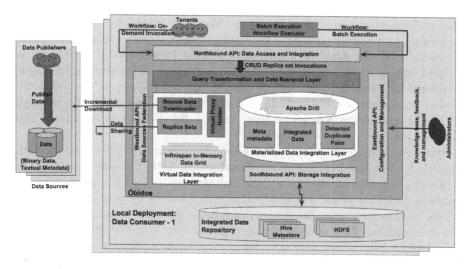

Fig. 2. *Óbidos* architecture: APIs, data integration layers, and the integrated data repository

Layer. When a data item or set that has not been loaded yet into the integrated data repository is accessed, *Óbidos* retrieves the virtual proxies of the relevant data in the virtual data integration layer. If the relevant virtual proxies do not exist or if they are insufficient to offer the results for the query, the data is loaded from the data sources. The integrated data is accessed through a distributed high performance **Query Transformation and Data Retrieval Layer** which has Apache Drill as its core query engine. This allows efficient queries to the data, partially or completely loaded into the integrated data repository.

Location, access, and public API details of the data sources are stored by the **Source Data Downloader** of *Óbidos*. Each data set that has been loaded by the tenant is identified by a replica set which functions as a pointer to the actual data that can resolve back to the search query. *Óbidos* bootstrapping follows a lazy loading approach. It retrieves metadata of the identified matching subsets as virtual proxies. The **Virtual Proxy Holder** offers an indexed and hierarchical data structure to map and store all the virtual proxies.

The components and data corresponding to virtual data integration layer reside in the In-Memory Data Grid to offer quick access to the critical and most accessed data, without requiring any disk access. The exact data matching the user query is fetched and stored in a highly scalable integrated data repository, along with the metametadata and detected duplicate pairs. The integrated data repository stores query outputs as well as data loaded by previous queries. Therefore, queries can be regarded as virtual data sets that can be reaccessed and shared (akin to the materialized view in traditional RDBMS).

The **Northbound API** is the user-facing API, invoked by the tenants to access and integrate the data. Invocation of the northbound API initiates various tenant-centric workflows that are discussed further in Sect. 4.1. Data sources and other data consumers are connected by the **Westbound API** of *Óbidos*. While the various data sources offer APIs providing access to their data, tenants create virtual data sets and share them with other tenants. Hence, *Óbidos* itself poses as a hybrid integrated data repository. The data download and sharing mechanisms are offered by the westbound API of *Óbidos*. The westbound API is indirectly invoked on-demand by the tenants through the northbound API or through the **batch execution workflow** that periodically checks and downloads the changesets of relevant data in batch.

System management and a feedback loop to improve the outcomes through a knowledge base are handled by the **Eastbound API**. The eastbound API is invoked by the administrators for data management tasks such as cleaning up orphaned data that is not referred by the replica sets or by periodic batch executions such as leveraging the detected duplicate pairs for subsequent iterative data integrations. The physical storage that functions as the integrated data repository for the materialized data integration layer is configured through the **Southbound API** by the administrators. *Óbidos* integrated data repository consists of HDFS and Hive metastore, deployed in a distributed cluster.

4   *Óbidos* Prototype for Medical Research Data

An *Óbidos* prototype has been implemented with Oracle Java 1.8 to integrate multimodal medical data from various heterogeneous data sources including The Cancer Imaging Archive (TCIA) [18], imaging data hosted in Amazon S3 buckets, medical images accessed through caMicroscope[5], clinical and imaging data hosted in local data sources including relational and NoSQL databases, and file system with files and directories along with CSV files as metadata.

4.1   RESTful Interfaces

Óbidos RESTful interface is designed as CRUD (Create, Retrieve, Update, and Delete) functions on replica sets. The tenant finds subsets of data from data sources by invoking **create** *replica set*. This creates a replica set and initiates the data loading workflow. When **retrieve** *replica set* is invoked, the relevant data is retrieved and *Óbidos* checks for updates from the data sources pointed by the replica set. Metadata of the replica set is compared against that of the data sources for any corruption or local changes. By running a near duplicate detection between, the local metadata representing the current state of the integrated data, and the metametadata stored at the time of data loading, the corrupted or deleted data is identified, redownloaded, and integrated.

The tenant updates an existing replica set to increase, decrease, or alter its scope, by invoking the **update** *replica set*. This may thus invoke parts of create and/or delete workflows, as new data may be loaded while existing parts of data may need to be removed. The tenant deletes existing replica sets by

---

[5] http://camicroscope.org.

invoking the **delete** *replica set*. The virtual data integration layer is updated immediately to avoid loading updates to the deleted replica sets. While the virtual data integration layer is multi-tenanted with each tenant having their own virtual isolated space, the integrated data repository is shared among them. Hence, before deleting, the data should be confirmed to be an orphan with no replica sets referring to them from any of the tenants. Deleting the data from the integrated data repository is designed as a separate process that is initiated by the administrators through the eastbound API. When the storage is abundantly available in a cluster, *Óbidos* advocates keeping orphan data in the integrated data repository rather than immediately initiating the cleanup process, and repeating it too frequently.

4.2 Data Representation

Óbidos virtual proxy holder consists of a few easy to index and space efficient data structures. *Óbidos* presents the replica sets internally in a minimal tree-like structure. Depending on how deep the structure is traversed, a replica set can be traced down to its virtual proxy by the virtue of its design as discussed below.

A multi-map (named 'tenant multi-map') lists the replica sets for each of the tenant. It stores replica sets belonging to each tenant, with the keys representing the tenant ID and each of the values representing one or more 'replica set maps'. As replica sets include pointers to data sets of interest from various data sources, each replica set map indicates the relevant data sources by each of the replica set, having replica set ID as the key and the list of data source names as the value. Each data source belonging to a replica set is internally represented by n maps. Each of the n maps represents one of the granularities in the data source (hence, $n = 4$ for TCIA), and a boolean array A_b of length n, each element $A_b[i]$ of the array representing the existence of a non-null entry in the ith map of granularity. Thus, the boolean flags in A_b indicate the existence (or lack thereof) of the data set in any granularity. If an entire level of granularity is included in the replica set by the tenant, the relevant flag is set to true.

Figure 3 illustrates the data representation of *Óbidos* and how it stores the replica sets and virtual proxies in a tree-like data structure. The maps representing each granularity resolve to store the identifier of the metadata or a virtual proxy. Hence details of data sets of interest from TCIA imaging metadata is stored in 4 maps - one each for collections, patients, studies, and series.

4.3 Implementation

Apache Velocity 1.7 [19] is leveraged to generate the application templates of the *Óbidos* web interface. Hadoop 2.7.2 and Hive 1.2.0 are used to store the integrated data and the metametadata respectively. Hive-jdbc package writes the metametadata into the Hive metastore. SparkJava 2.5[6] compact Java-based web framework is leveraged to expose the *Óbidos* APIs as RESTful services. The APIs

---

[6] sparkjava.com/.

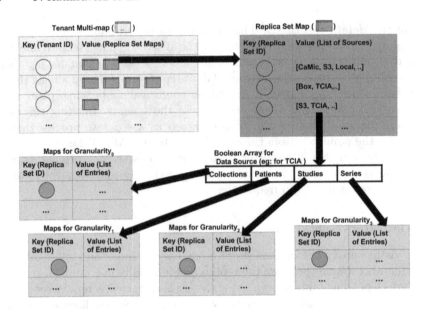

Fig. 3. Data representation: *Óbidos* virtual data integration layer

are managed and made available to the relevant users through API gateways. API Umbrella is deployed as the default API gateway. Embedded Tomcat 7.0.34 is used to deploy *Óbidos* with minimal deployment and operational overhead. The Embedded Tomcat-based web application exposes *Óbidos* APIs and offers a simple web-based user interface to the data consumers to access the data sources, and create, modify, and share replica sets. Infinispan 8.2.2 is used as the In-Memory Data Grid where its distributed streams support distributed execution of *Óbidos* workflows across the *Óbidos* clustered deployment. The virtual proxy holder and the replica sets are represented by instances of Infinispan Cache class, which is a Java implementation of distributed HashMap.

Drill 1.7.0 is exploited for the SQL queries on the integrated data repository, with drill-jdbc offering JDBC API to interconnect with Drill from the Query Transformation and Data Retrieval Layer. Drill cannot query nested arrays inside data in unstructured formats such as JSON (or NoSQL data sources generated from them), unless its *flatten* function is used. However, *flatten* alters the data structure and flattens the objects in the query results to make multiple entries instead of a single object with a nested array. This may be unfit for the requirements of a data consumer. We also observed *flatten* to be slow on large nested arrays. Hence, when this situation is expected, *Óbidos* deployment is configured by the administrators to either (i) avoid Drill and query the data integration layer directly from the query transformation and data retrieval layer, or (ii) configure the southbound such that the nested arrays are converted into an array of maps at the time of data loading into the integrated data repository.

5   Evaluation

Óbidos has been benchmarked against implementations of eager ETL and lazy ETL approaches, using microbenchmarks derived from medical research queries on cancer imaging and clinical data. While *Óbidos* can integrate textual and binary (based on their metadata) data efficiently, we limit our focus to DICOM images to benchmark against the traditional data loading and sharing approaches in a fair manner.

The DICOM imaging data used in the evaluation consists of collections of various volume, as shown by Fig. 4. The data consists of large scale binary images (in the scale of a few thousands GB, up to 10 000 GB) along with a smaller scale textual metadata (in the range of MBs).

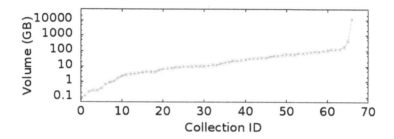

Fig. 4. Evaluated DICOM imaging collections (sorted by total volume)

Various granularities of DICOM are leveraged in the evaluations. Figure 5 illustrates the number of patients, studies, series, and images in each of the collection. Collections are sorted according to their total volume. Each collection consists of multiple patients; each patient has one or more studies; each study has one or more series; and each series has multiple images. Figure 5 indicates that the total volume of a collection does not necessarily reflect the number of entries in it.

5.1   Loading and Querying Data

Óbidos was benchmarked for its performance and efficiency in loading the data and average query time from the integrated data repository. Initially, to avoid the influence of bandwidth or transfer speed in the evaluation, data sources were replicated to a local cluster in the first two iterations, illustrated by Figs. 6 and 7. Figure 6 shows the loading time for different total volume of data sources for the same data sets of interest of the tenant. Figure 7 shows the time taken for the same total volume of data sources while increasing the number of studies of interest.

Finally, data was loaded from the remote data sources through their web service APIs, to evaluate the influence of data downloading and bandwidth consumption associated with it. The total volume of data was changed, while keeping the data sets of interest constant. Figure 8 shows the time taken for *Óbidos*

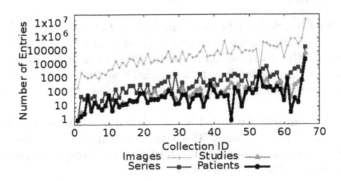

Fig. 5. Various entries in evaluated collections (sorted by total volume)

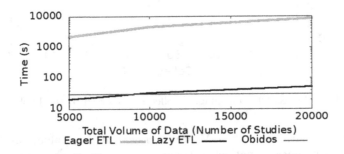

Fig. 6. Data load time: change in total volume of data sources (same query and same data set of interest)

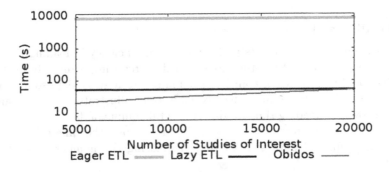

Fig. 7. Data load time: varying number of studies of interest (same query and constant total data volume)

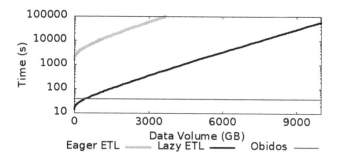

Fig. 8. Load time from the remote data sources

and the other approaches. Eager ETL performed poor as binary data had to be downloaded over the network. Lazy ETL too performed slow for large volumes as it must eagerly load the metadata (which itself grows with scale) over the network.

Since *Óbidos* loads the metadata of only the data sets of interest, the loading time remains constant for the experiments as illustrated by Figs. 6 and 8, where it increases with the number of studies of interest in the experiment depicted by Fig. 7. As only the data sets of interest are loaded, *Óbidos* uses bandwidth conservatively, loading no irrelevant data or metadata.

Query completion time depends on the number of entries in the queried data rather than the size of the entire integrated data repository. Hence, varying amounts of data, measured by the number of studies, were queried. The query completion time is depicted in Fig. 9. The sample queries used are made sure to be satisfied by the data that have already been loaded, in case of lazy ETL and *Óbidos*. In the sample deployments, eager ETL and lazy ETL offered similar performance for various queries. *Óbidos* showed a speedup compared to the lazy ETL, which can be attributed to the efficiency of the query transformation and data retrieval layer. The unstructured data in HDFS was very efficiently queried as in a relational database through the distributed query execution of Drill with its SQL support for NoSQL data sources. Eager ETL would outperform both lazy ETL and *Óbidos* for queries that access data not yet loaded (in lazy ETL and *Óbidos*), as eager ETL would have constructed an entire data warehouse beforehand. *Óbidos* attempts to minimize this by leveraging the virtual proxies in addition to the metadata. Furthermore, with the domain knowledge of the medical data researcher, the relevant subsets of data are loaded timely. The time required to construct such a data warehouse would prevent any benefits of eager loading from being prominent. If data is also not loaded beforehand in eager ETL, it would consume much longer to construct the entire data warehouse before starting the query. Moreover, loading everything beforehand is irrelevant and impractical or even impossible for study-specific researches due to the scale and distribution of sources.

Overall, in all the relevant use cases, lazy ETL and *Óbidos* significantly outperformed eager ETL as the need to build a complete data warehouse is avoided

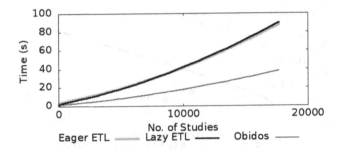

Fig. 9. Query completion time for the integrated data repository

in them. As *Óbidos* loads only the subset of metadata, and does not even eagerly load the metadata unlike eager ETL, for large volumes, *Óbidos* also significantly outperformed lazy ETL. While eager ETL performs better for data that has not been loaded prior by *Óbidos* and lazy ETL, *Óbidos* assumes that the researcher is aware of the data to load the data sets of interest appropriately.

5.2 Sharing Efficiency of Medical Research Data

Various image series of an average uniform size are shared between tenants inside an *Óbidos* deployment and between remote deployments. Figure 10 benchmarks the data shared in these *Óbidos* data sharing approaches against the typical binary data transfers for its bandwidth efficiency. Inside *Óbidos* deployment, regardless of the size of data sets to be shared, only an identifier to the replica set, the replica set ID, is shared. The ID is interpreted directly by the receiving tenant from *Óbidos*. Hence, the shared data is of a constant size. When data is shared between multiple *Óbidos* deployments, replica sets are shared. Replica sets are minimal in size as pointers to actual data. However, they grow linearly when more data of same granularity is shared. Minimal data was shared in both cases compared to sharing actual data.

This also avoids the need for manually sharing the locations of the data sets, which is an alternative bandwidth-efficient approach to sharing the integrated data. As the pointers are shared, no actual data is copied and shared. This enables data sharing with zero redundancy.

6 Related Work

Virtual data integration and service-based data integration approaches are two major motivations for *Óbidos*. *Óbidos* leverages both approaches together for an enhanced functionality and performance.

Service-Based Data Integration: OGSA-DAI (Open Grid Services Architecture - Data Access and Integration) [20] facilitates federation and management of various data sources through its web service interface. The Vienna Cloud Environment (VCE) [21] offers service-based data integration of clinical trials and

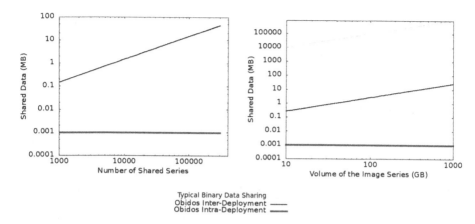

Fig. 10. Actual data shared in *Óbidos* data sharing use cases vs. in regular binary data sharing

consolidates data from distributed sources. VCE offers data services to query individual data sources and to provide an integrated schema atop the individual data sets. This is similar to *Óbidos*, though *Óbidos* leverages virtual data integration for an efficient federation and sharing of data and avoids replicating data excessively.

EUDAT [22] is a platform to store, share, and access multidisciplinary scientific research data. EUDAT hosts a service-based data access feature B2FIND [23], and a sharing feature B2SHARE [24]. When researchers access these cross-disciplinary research data sources, they already know which of the repositories they are interested in, or can find them by the search feature. Similar to the motivation of *Óbidos*, loading the entire data from all the sources is irrelevant in EUDAT. Hence, choosing and loading certain sets of data is supported by these service-based data access platforms. *Óbidos* can be leveraged to load related cross-disciplinary data from the eScience data sources such as EUDAT.

Heterogeneity in data and sources, and the distribution of data sources, have imposed various challenges to the data integration. Multi-agent systems have been proposed for data integration in wireless sensor networks [25]. *Óbidos* approach can be used in conjunction with multi-agent systems where metadata can be loaded instead of an eager loading of data, while still caching the data loaded by the distributed agents.

Lazy ETL: Lazy ETL [6] demonstrates how metadata can be efficiently used for study-specific queries without actually constructing an entire data warehouse beforehand, by using files in SEED [26] standard format for seismological research. The hierarchical structure and metadata of SEED is similar to that of DICOM medical imaging data files that are accessed by the *Óbidos* prototype. Thus, we note that while we prototype *Óbidos* for medical research, the approach is applicable to various research and application domains.

LigDB [27] is similar to *Óbidos* as both focus on a query-based integration approach as opposed to having a data warehouse constructed as the first step, and it efficiently handles unstructured data with no schema. However, *Óbidos* differs as it indeed has a scalable storage, and does not periodically evict the stored data unlike LigDB.

Medical Data Integration: Leveraging Hadoop ecosystem for management and integration of medical data is not entirely new, and our choices are indeed motivated by previous work [28]. However, the existing approaches failed to extend the scalable architecture offered by Hadoop and the other big data platforms to create an index to the unstructured integrated data, manage the data in-memory for quicker data manipulations, and share results and data sets efficiently with peers. *Óbidos* attempts to address these shortcomings with its novel approach and architecture, combining service-based big data integration with virtual data integration, designed for research collaboration.

Our previous work, Data Café [29] offers a dynamic warehousing platform for medical research data integration. It lets the medical big data researchers to construct a data warehouse, integrating data from heterogeneous data sources, consisting of unstructured or semi-structured data, even without knowing the data schema a priori. *Óbidos* leverages and extends Data Café in materialized data integration layer to store the loaded binary data. *Óbidos* also shares the (Apache Hive and Drill-based) data query architecture of Data Café to efficiently index and query the integrated data repository.

7   Conclusion and Future Work

Óbidos offers service-based big data integration to ensure fast and resource efficient data analysis. It stores the loaded binary data in a distributed and scalable integrated data repository, and the metadata representing a larger set of data in an in-memory cluster. By implementing and evaluating *Óbidos* for medical research data, we demonstrated how *Óbidos* offers a hybrid data integration solution for big data, and lets the researchers share data sets with zero redundancy.

We built our case on the reality that data sources are proliferating, and cross-disciplinary researches such as medical data research often require access and integration of data sets spanning across the multiple data sources in the Internet. We further presented how a service-based data integration and access approach fits well for the requirement to have a protected access of sensitive data by various researchers. We leveraged the respective APIs offered by the data sources in consuming and loading the data, while offering our own service-based access to the *Óbidos* integrated data repository. We further envisioned that various distributed deployments of *Óbidos* will be able to collaborate and coordinate to construct and share study-specific integrated data sets internally and between one another.

As a future work, we aim to deploy *Óbidos* approach to consume data from various research repositories such as EUDAT to find and integrate research data.

Thus, we will be able to conduct a usability evaluation of *Óbidos* based on different use cases and data sources. We also propose to leverage the network proximity among the data sources and the data consumer deployments for efficient data integration and sharing, in the future work. Thus, we aim to build virtual distributed data warehouses - data partially replicated and shared across various research institutes.

Acknowledgements. This work was supported by NCI U01 [1U01CA187013-01], Resources for development and validation of Radiomic Analyses & Adaptive Therapy, Fred Prior, Ashish Sharma (UAMS, Emory), national funds through Fundação para a Ciência e a Tecnologia with reference UID/CEC/50021/2013, PTDC/EEI-SCR/6945/2014, a Google Summer of Code project, and a PhD grant offered by the Erasmus Mundus Joint Doctorate in Distributed Computing (EMJD-DC).

References

1. Lee, G., Doyle, S., Monaco, J., Madabhushi, A., Feldman, M.D., Master, S.R., Tomaszewski, J.E.: A knowledge representation framework for integration, classification of multi-scale imaging and non-imaging data: preliminary results in predicting prostate cancer recurrence by fusing mass spectrometry and histology. In: 2009 IEEE International Symposium on Biomedical Imaging: From Nano to Macro, pp. 77–80. IEEE (2009)
2. Huang, Z.: Data Integration For Urban Transport Planning. Citeseer (2003)
3. Sujansky, W.: Heterogeneous database integration in biomedicine. J. Biomed. Inform. **34**(4), 285–298 (2001)
4. Mildenberger, P., Eichelberg, M., Martin, E.: Introduction to the dicom standard. Eur. Radiol. **12**(4), 920–927 (2002)
5. Whitcher, B., Schmid, V.J., Thornton, A.: Working with the DICOM and NIFTI data standards in R. J. Stat. Softw. **44**(6), 1–28 (2011)
6. Kargín, Y., Ivanova, M., Zhang, Y., Manegold, S., Kersten, M.: Lazy ETL in action: ETL technology dates scientific data. Proc. VLDB Endow. **6**(12), 1286–1289 (2013)
7. Dong, X.L., Srivastava, D.: Big data integration. In: 2013 IEEE 29th International Conference on Data Engineering (ICDE), pp. 1245–1248. IEEE (2013)
8. Rustagi, A.: Parallel processing for ETL processes. US Patent App. 11/682,815 (2007)
9. Porter, D.L., Swanholm, D.E.: Distributed extract, transfer, and load (ETL) computer method. US Patent 7,051,334 (2006)
10. Rimal, B.P., Choi, E., Lumb, I.: A taxonomy and survey of cloud computing systems. In: INC, IMS and IDC, pp. 44–51 (2009)
11. Kathiravelu, P., Galhardas, H., Veiga, L.: $\partial u \partial u$ multi-tenanted framework: distributed near duplicate detection for big data. In: Debruyne, C., et al. (eds.) OTM 2015. LNCS, vol. 9415, pp. 237–256. Springer, Cham (2015). doi:10.1007/978-3-319-26148-5_14
12. White, T.: Hadoop: The Definitive Guide. O'Reilly Media Inc., Sebastopol (2012)
13. Marchioni, F., Surtani, M.: Infinispan Data Grid Platform. Packt Publishing Ltd., Birmingham (2012)
14. Thusoo, A., Sarma, J.S., Jain, N., Shao, Z., Chakka, P., Anthony, S., Liu, H., Wyckoff, P., Murthy, R.: Hive: a warehousing solution over a map-reduce framework. Proc. VLDB Endow. **2**(2), 1626–1629 (2009)

15. Hausenblas, M., Nadeau, J.: Apache drill: interactive ad-hoc analysis at scale. Big Data **1**(2), 100–104 (2013)
16. Veiga, L., Ferreira, P.: Incremental replication for mobility support in OBIWAN. In: Proceedings of the 22nd International Conference on Distributed Computing Systems, pp. 249–256. IEEE (2002)
17. Xiao, C., Wang, W., Lin, X., Yu, J.X., Wang, G.: Efficient similarity joins for near-duplicate detection. ACM Trans. Database Syst. (TODS) **36**(3), 15 (2011)
18. Clark, K., Vendt, B., Smith, K., Freymann, J., Kirby, J., Koppel, P., Moore, S., Phillips, S., Maffitt, D., Pringle, M., et al.: The cancer imaging archive (TCIA): maintaining and operating a public information repository. J. Digit. Imaging **26**(6), 1045–1057 (2013)
19. Gradecki, J.D., Cole, J.: Mastering Apache Velocity. Wiley (2003)
20. Antonioletti, M., Atkinson, M., Baxter, R., Borley, A., Chue Hong, N.P., Collins, B., Hardman, N., Hume, A.C., Knox, A., Jackson, M., et al.: The design and implementation of grid database services in OGSA-DAI. Concurr. Comput. Pract. Exp. **17**(2–4), 357–376 (2005)
21. Borckholder, C., Heinzel, A., Kaniovskyi, Y., Benkner, S., Lukas, A., Mayer, B.: A generic, service-based data integration framework applied to linking drugs & clinical trials. Procedia Comput. Sci. **23**, 24–35 (2013)
22. Lecarpentier, D., Wittenburg, P., Elbers, W., Michelini, A., Kanso, R., Coveney, P., Baxter, R.: EUDAT: a new cross-disciplinary data infrastructure for science. Int. J. Digit. Curation **8**(1), 279–287 (2013)
23. Widmann, H., Thiemann, H.: EUDAT B2FIND: a cross-discipline metadata service and discovery portal. In: EGU General Assembly Conference Abstracts, vol. 18, p. 8562 (2016)
24. Ardestani, S.B., Håkansson, C.J., Laure, E., Livenson, I., Stranák, P., Dima, E., Blommesteijn, D., van de Sanden, M.: B2SHARE: an open eScience data sharing platform. In: 2015 IEEE 11th International Conference on e-Science (e-Science), pp. 448–453. IEEE (2015)
25. Qi, H., Iyengar, S., Chakrabarty, K.: Multiresolution data integration using mobile agents in distributed sensor networks. IEEE Trans. Syst. Man Cybern. Part C (Appl. Rev.) **31**(3), 383–391 (2001)
26. Ahern, T., Casey, R., Barnes, D., Benson, R., Knight, T.: Seed standard for the exchange of earthquake data reference manual format version 2.4. Incorporated Research Institutions for Seismology (IRIS), Seattle (2007)
27. Milchevski, E., Michel, S.: ligDB-online query processing without (almost) any storage. In: EDBT, pp. 683–688 (2015)
28. Lyu, D.M., Tian, Y., Wang, Y., Tong, D.Y., Yin, W.W., Li, J.S.: Design and implementation of clinical data integration and management system based on Hadoop platform. In: 2015 7th International Conference on Information Technology in Medicine and Education (ITME), pp. 76–79. IEEE (2015)
29. Kathiravelu, P., Sharma, A.: A dynamic data warehousing platform for creating and accessing biomedical data lakes. In: Wang, F., Yao, L., Luo, G. (eds.) DMAH 2016. LNCS, vol. 10186, pp. 101–120. Springer, Cham (2017). doi:10.1007/978-3-319-57741-8_7

CHIPS – A Service for Collecting, Organizing, Processing, and Sharing Medical Image Data in the Cloud

Rudolph Pienaar[1,2]([✉]), Ata Turk[3], Jorge Bernal-Rusiel[1], Nicolas Rannou[5],
Daniel Haehn[4], P. Ellen Grant[1,2], and Orran Krieger[3]

[1] Boston Children's Hospital, Boston, MA 02115, USA
`rudolph.pienaar@childrens.harvard.edu`
[2] Harvard Medical School, Boston, MA 02115, USA
[3] Boston University, Boston, MA 02115, USA
[4] Harvard School of Engineering and Applied Sciences, Cambridge, MA 02138, USA,
[5] Eunate Technology S.L., Sopela, Spain

Abstract. Web browsers are increasingly used as middleware platforms offering a central access point for service provision. Using backend containerization, RESTful APIs, and distributed computing allows for complex systems to be realized that address the needs of modern compute intense environments. In this paper, we present a web-based medical image data and information management software platform called *CHIPS* (*C*loud *H*ealthcare *I*mage *P*rocessing *S*ervice). This cloud-based services allows for authenticated and secure retrieval of medical image data from resources typically found in hospitals, organizes and presents information in a modern feed-like interface, provides access to a growing library of plugins that process these data, allows for easy data sharing between users and provides powerful 3D visualization and real-time collaboration. Image processing is orchestrated across additional cloud-based resources using containerization technologies.

Keywords: Web-based neuroimaging · Big-data · Applied containerization · Web-based collaborative visualization · Real-time collaboration · HTML5 · Web services · Telemedicine · Cloud-storage

1 Introduction

Modern web browsers are becoming powerful platforms for advanced application development [6,9]. New advances in core web application technologies such as the modern web browsers' universal support of ECMAScript 5 (and 6) [10], CSS3 and HTML5 APIs have made it much more feasible to implement powerful middle-ware platforms for data management and powerful graphical rendering, as well as real-time communication purely in client-side JavaScript [2,12]. The last decade has seen a slow, but steady, shift to fully distributed solutions using web-standards [4,11,15,17], closely tracked by expressiveness of the JavaScript

© Springer International Publishing AG 2017
E. Begoli et al. (Eds.): DMAH 2017, LNCS 10494, pp. 29–35, 2017.
DOI: 10.1007/978-3-319-67186-4_3

programming language. Web-based solutions are especially appealing as they do not require the installation of any client-side software other than a standard web browser which enhances accessibility and usability.

Unrelated to rise of web-technologies, a new emerging trend is the rapid adoption of containerization technologies. These have enabled the concept of *compute* portability in a similar sense to *data* portability. Just as data can be moved from place to place, containerization allows for operations on that data to also be moved from place to place.

To our knowledge, no web-based platform currently exists that provides data *and* compute agnostic services (some services, such as CBRAIN [15] and LONI [14] provide conceptually similar approaches, but do not have deep connectivity to typical hospital database repositories), in particular collection, management, and real-time sharing of medical data, as well as access to pipelines that process that data. In this paper, we introduce *CHIPS* (**C**loud **H**ealthcare **I**mage **P**rocessing **S**ervice). *CHIPS* is a novel web-based medical data storage and data processing workflow service that provides strict data security while also facilitating secure, real-time interactive collaboration over the Internet and internal Intranets.

CHIPS is able to seamlessly collect data from typical sources found in hospitals (such as Picture Archive and Communications Systems, PACS) and easily export to approved cloud storage. *CHIPS* not only manages data collection and organization, but it also provides a large (and expanding) library of pipelines to analyze imported data, and the containerized compute can execute in a large variety of remote resources. *CHIPS* provides for persistent record and management of activity in *feeds* as well as for powerful visualization of data. In particular, it makes use of the popular XTK toolkit which was also developed by our team at the Fetal-Neonatal Neuroimaging and Developmental Science Center, Boston Childrens Hospital[1] for the in-browser rendering and visualization of medical image data and can be freely downloaded from the web[2] [8].

2   Architectural Overview

2.1   Scope

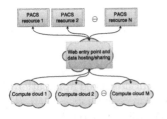

Fig. 1. *CHIPS* connects multiple input PACS sources to multiple "cloud" compute nodes.

The creation of *CHIPS* has been motivated by both clinical and research needs. On the clinical side, *CHIPS* was built to provide clinicians with easy access to large amounts of data (especially from hospital image databases like Picture Archive and Communications Systems – PACS), to provide for powerful collaboration, and to allow for easy access to a library of analysis processes or pipelines. On the research side, *CHIPS* was designed to allow computational

---

[1] http://fnndsc.babymri.org.
[2] http://goxtk.com.

researchers to test and develop new algorithms for image processing across heterogeneous platforms, while allowing life science researchers to focus on their research protocols and data processing, without needing to spend time on the minutiae of performing data analysis.

The system design is highly distributed, as shown in Fig. 1, which shows a *CHIPS* deployment connected to multiple input sources and multiple compute sources. Though the figure suggests a single, discrete central point, components of *CHIPS* do reside on each input (PACS) and compute location.

2.2   Distributed Component Design

Architecturally *CHIPS* is not a single monolithic system, but a distributed collection of interconnected components, including a front-end webserver and web-based UI; a core RESTful back-end central server that provides access to all data, feeds, users, etc.; a DICOM/PACS interface; a set of independent RESTful microservices that handle inter-network data IO and also remote process management, and a core cloud-based computational platform that orchestrates offloading of image processing pipelines to some remote cloud-based compute – see Fig. 2.

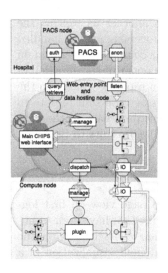

The top the red box of Fig. 2 contains the *PACS node* and represents the Hospital image data repository. The second blue box, labeled *Web-entry point and data hosting node* contains the main *CHIPS* backend and is presented as being in a "cloud" (i.e. some resource that is accessible from the Internet). Finally, the bottom yellow box is shown on a separate "cloud" to emphasize that it is topologically distinct from the *Web-entry point*.

The logical relationships between data (represented as the rectangles with a tree structure) and compute elements denoted by the named hexagons is shown by either data connectors (thick blue arrows) or control connections (single line arrows). In the syntax of the diagram, the stylized cloud icon touching some of the boxes denotes that these compute elements are controlled by a REST API, while the sphere icon denotes web-access.

Fig. 2. The internal *CHIPS* logical architecture. (Color figure online)

An remote compute is denoted by `plugin`, which is controlled by a `manage` component. In the most abstract sense, the `plugin` processes an input data structure, and outputs a transformed data structure (the two tree graphs as shown). File transfer between the data cloud and compute cloud is performed by the file `IO` handler component. A `query/retrieve` process in the data cloud connects to an authentication process, `auth` in the Hospital network, while on-the-fly anonymization of DICOM images is handled by process anonymizer anon. Finally the dispatcher is a component that determines what compute node (or cloud) is best suited for the data analysis at hand. The circle icon attached to the manage and plugin icons implies the attached process and can provide real-time feedback information to other software agents about the controlled process via its own REST interface.

2.3   Pervasive Containerization

CHIPS is designed as a distributed system, and the underlying components are containerized (currently) using docker[3]. In Fig. 2, the *Main CHIPS web interface* and associated backend database is housed within a single container[4]. Input data and processed results are accessible in the hosting node and volume mapped as appropriate to this back end. Other components of *CHIPS* in the web-entry node are similarly containerized. This includes the manage[5] block, which is responsible for spawning processes on the underlying system. Not only does manage provide the means to start and stop processes, but it also tracks the execution state, termination state, and standard output/error streams of the process. The manage component has a REST interface through which clients can start/stop and query processes.

Also containerized is the IO[6] component that can transfer entire directory trees across network boundaries from one system to another as well as the dispatch[7] component that can orchestrate multiple processing jobs as handled by manage. The plugin container houses the particular compute to perform on a given set of data, and is spawned by the manage component under direction of the dispatch. Since the compute typically occurs on a separate system to the data hosting node, the IO containers perform the necessary transmit of data to this compute system, as well as the retrieve of resultant data back to the data node, allowing the web container to present (and visualize) results to the user.

3   UI Considerations

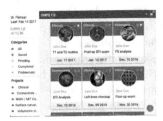

Fig. 3. *CHIPS* home page with a "cards" organization.

Figure 3 shows the home page view on first logging into the system. Studies that have been "sent" to *CHIPS* appear in their own "cards" on the user's home page with a small visualization of a represented image set of the study. Various control on this home page allow users to organize/tag "cards" in specific projects (or folders), remove cards, bookmark for easy access, etc. New cards can be generated by clicking on the ⊕ icon and

[3] https://www.docker.com.
[4] https://github.com/FNNDSC/ChRIS_ultron_backEnd.
[5] https://github.com/FNNDSC/pman.
[6] https://github.com/FNNDSC/pfioh.
[7] https://github.com/FNNDSC/swarm.

choosing an activity (such as PACS Query/Retrieve), and any card can be seamlessly shared with other users of the system.

On selecting a given feed, the core image data in that feed is visualized in a rich, web-based viewer – see Fig. 4. Various tabs and elements of the feed view provide different perspectives on the data, and also provide the ability to annotate notes, or add comments. As in the feed view, a ⊕ icon is also present, and if selected, opens a ribbon of "plugins" (or "apps") to run on the data contained in the feed. For example, certain plugins might perform a surface reconstruction of the brain surface with tissue segmentation (for example, a FreeSurfer plugin).

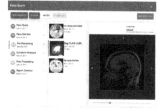

The interface semantics within a feed are straightforward: a user clicks on the feed and enters the top level data view. Once a plugin from the ⊕ is applied, the feed data is processed accordingly. When the plugin is completed, its output files are also organized in the feed in a logical tree view (accessible via the left "Data" tab) in a manner akin to an email thread. In this manner, the thread of execution from data → plugin → data is defined – in effect building a workflow.

Fig. 4. Visualizing pulled and processed data.

Any image visualized can also be shared in real-time using collaboration features built into the viewer library and leveraging the Google Drive API and Google Realtime API [2].

4 Big Data Infrastructure

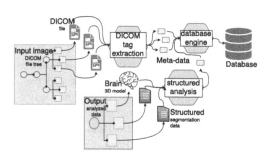

Fig. 5. Big data pre-processing. (Color figure online)

An important component of *CHIPS* lies in creating a foundation suitable for future support of "data mining". Recently, the term *Big Data* has come into common parlance, especially in the context of informatics [7,13,16]. Despite the term and the use of *Big*, the concept often refers to the use of predictive analytics and other advanced data analytics tools that extract meaning from sets of data and does not necessarily to the particular size of the data set.

In healthcare, big data analytics has impacted the field in very specific areas such as clinical risk intervention, waste and care variability reduction, and automated reporting. However, as a field, biomedical imaging has not especially benefited from big data approaches due to the unstructured nature of image data, complexity of results from analysis in terms of data formats (again usually unstructured), simple quality issues such as noise in image acquisitions, etc.

CHIPS constructs a framework to allow big data methods to be used in this image space. Consider that the incoming source data to *CHIPS* are DICOM images that by their nature contain a large amount of meta information, most of which is non PHI and will be left unchanged by the anonymization processes. Information about the scanning/imaging protocol, acquisition parameters, as well as certain non-PHI demographics such as patient sex and age can be meaningfully databased. Moreover, the application of an analysis pipeline to an image data-set can in turn result in large amounts of meaningful data that can be data-based and associated with the incoming source data. For example, FreeSurfer, which is dockerized as a plugin in the *CHIPS* system produces volumetric segmentations and surface reconstructions on raw input MRI T1 weighted data [1,3,5].

In Fig. 5 input raw DICOM (purple block) and output processed data from the DICOMs (green block) are shown. A `DICOM tag extraction` process removes the image meta data and associates this information with the particular image record. DICOM data is regularly formatted and easily extracted. Importantly, for the output data, and assuming the output data is a 3D surface reconstruction and tables of brain parcellation volume values, a `structured analysis` process regularizes all this information into meta data that will be added to the space of data pertaining to this image record. This processing will lay the ground work on which data analytics can explore and mine for relations between (for example) input acquisition parameters and pipeline output results, or simply mine across output results for hidden trends in data trajectories (for example volumetric changes with age or sex).

5   Conclusion and Future Directions

CHIPS is a distributed system that provides a single, cloud-based, access point to a large family of services. These include: (a) accessing medical image data securely from participating institutions with authenticated access and built-in anonymization of collected image data; (b) organizing collected data in a modern UI that allows for easy data management and sharing; (c) performing processing on images by dispatching data to remote clouds and controlling/managing remote execution on these resources; (d) powerful real-time collaboration on images using secure third party services (such as the Google RealTime API); and intuitively constructing medical image processing workflows. *CHIPS* is not only a medical data management system, but strives to improve the quality of healthcare by allowing clinical users the ability to easily perform value added processing and sharing of data and information. Current and future directions for *CHIPS* include facilitating the construction of big-data frameworks and allowing for users to simply construct experiments for data analytics and various machine learning pipelines.

All analysis and development conducted by the *CHIPS* system at the Boston Children's Hospital was conducted under relevant Institutional Review Board approval, which governed access to image data and controlled the scope of sharing of such data.

References

1. FreeSurfer. http://surfer.nmr.mgh.harvard.edu/
2. Bernal-Rusiel, J.L., Rannou, N., Gollub, R., Pieper, S., Murphy, S., Robertson, R., Grant, P.E., Pienaar, R.: Reusable client-side javascript modules for immersive web-based real-time collaborative neuroimage visualization. Front. Neuroinformatics **11**, 32 (2017)
3. Dale, A.M., Fischl, B., Sereno, M.I.: Cortical Surface-Based Analysis - I. Segmentation and Surface Reconstruction. Neuroimage **9**, 179–194 (1999)
4. Eckersley, P., Egan, G.F., De Schutter, E., Yiyuan, T., Novak, M., Sebesta, V., Matthiessen, L., Jaaskelainen, I.P., Ruotsalainen, U., Herz, A.V., et al.: Neuroscience data and tool sharing. Neuroinformatics **1**(2), 149–165 (2003)
5. Fischl, B., Sereno, M.I., Dale, A.M.: Cortical surface-based analysis II: Inflation, flattening, and a surface-based coordinate system. NeuroImage **9**, 195–207 (1999)
6. Ginsburg, D., Gerhard, S., Calle, J.E.C., Pienaar, R.: Realtime visualization of the connectome in the browser using webgl. Front. Neuroinformatics (2011)
7. Greene, C.S., Tan, J., Ung, M., Moore, J.H., Cheng, C.: Big data bioinformatics. J. Cell. Physiol. **229**(12), 1896–1900 (2014). http://dx.doi.org/10.1002/jcp.24662
8. Haehn, D., Rannou, N., Ahtam, B., Grant, E., Pienaar, R.: Neuroimaging in the browser using the x toolkit. In: Frontiers in Neuroinformatics Conference Abstract: 5th INCF Congress of Neuroinformatics, Munich (2014)
9. Haehn, D., Rannou, N., Grant, P.E., Pienaar, R.: Slice: drop: collaborative medical imaging in the browser. In: ACM SIGGRAPH 2013 Computer Animation Festival, SIGGRAPH 2013, p. 1. ACM, New York (2013). http://doi.acm.org/10.1145/2503541.2503645
10. Khan, F., Foley-Bourgon, V., Kathrotia, S., Lavoie, E., Hendren, L.: Using Javascript and WebCL for numerical computations: a comparative study of native and web technologies. In: ACM SIGPLAN Notices, vol. 50, pp. 91–102. ACM (2014)
11. Millan, J., Yunda, L.: An open-access web-based medical image atlas for collaborative medical image sharing, processing, web semantic searching and analysis with uses in medical training, research and second opinion of cases. Nova **12**(22), 143–150 (2014)
12. Mwalongo, F., Krone, M., Reina, G., Ertl, T.: State-of-the-art report in web-based visualization. In: Computer Graphics Forum, vol. 35, pp. 553–575. Wiley Online Library (2016)
13. Provost, F., Fawcett, T.: Data science and its relationship to big data and data-driven decision making. Big Data **1**(1), 51–59 (2013). http://dx.doi.org/10.1089/big.2013.1508
14. Rex, D.E., Ma, J.Q., Toga, A.W.: The LONI pipeline processing environment. Neuroimage **19**(3), 1033–1048 (2003). http://www.hubmed.org/display.cgi?uids=12880830
15. Sherif, T., Rioux, P., Rousseau, M.E., Kassis, N., Beck, N., Adalat, R., Das, S., Glatard, T., Evans, A.C.: Cbrain: a web-based, distributed computing platform for collaborative neuroimaging research. Front. Neuroinformatics **8**, 54 (2014)
16. Swan, M.: The quantified self: Fundamental disruption in big data science and biological discovery. Big Data **1**(2), 85–99 (2013). http://dx.doi.org/10.1089/big.2012.0002
17. Wood, D., King, M., Landis, D., Courtney, W., Wang, R., Kelly, R., Turner, J.A., Calhoun, V.D.: Harnessing modern web application technology to create intuitive and efficient data visualization and sharing tools. Front. Neuroinformatics **8**, 71 (2014)

High Performance Merging of Massive Data from Genome-Wide Association Studies

Xiaobo Sun[1], Fusheng Wang[2], and Zhaohui Qin[3]([✉])

[1] Department of Computer Sciences, Emory University, Atlanta, USA
[2] Department of Computer Sciences, New York State University,
Stony Brook, USA
[3] Department of Biostatistics, Emory University, Atlanta, USA
Zhaohui.qin@emory.edu

Abstract. The traditional data processing methods working on single computer show less scalability and efficiency for performing ordered full-outer-joining, on merging large number of individual Genome-Wide Associations Studies (GWAS) data. Although the emerging of big data platforms such as Hadoop and Spark shed lights on this problem, the inefficiency of keeping data in total-sorted order as well as the workload imbalance problem limit their performance. In this study, we designed and compared three new methodologies based on MapReduce, HBase and Spark respectively, to merge hundreds of individuals VCF files on their Single Nucleotide Polymorphism (SNP) location into a single TPED file. Our methodologies overcame the limitations stated above and considerably improved the performance with good scalability on input size and computing resources.

Keywords: Genome-Wide Association Studies (GWAS) · Variant Call Format (VCF) · TPED · Total order full-outer-merging · MapReduce · HBase · Spark · Scalability

1 Introduction

Studies like Genome Wide Association Studies (GWASs) have produced massive amount of data. Platforms including MapReduce, HBase and the recent Spark offer the ability to efficiently process such Big Data (Chang et al. 2008; Dean and Ghemawat 2008; Massie et al. 2013; Zaharia et al. 2012). In the GWAS domain, single machine based software such as VCFTools (Danecek et al. 2011), PLINK (Purcell et al. 2007) are popular. Along with ever-increasing scale of GWAS data, there comes the needs of converting data between formats of these tools efficiently.

2 Problem Definition

In this study, whole genome SNP data of 690 individuals generated from Illumina's BaseSpace software were collected as VCFv4.1 files and merged as a single TPED file by their SNP locations.

© Springer International Publishing AG 2017
E. Begoli et al. (Eds.): DMAH 2017, LNCS 10494, pp. 36–40, 2017.
DOI: 10.1007/978-3-319-67186-4_4

Each file is of approximate 4.5 million rows corresponding to individual-specific genetic variant positions, with a size of about 300 MB. Each record row also contains a genotype associated with its SNP location. Therefore, this merging process is equivalent to an ordered full-out-joining problem. Software tools such as VCFTools offer utilities to do the merging and converting. However, most of these tools adopt the multiway-merge method to keep the order of merged results, which literally makes it single-threaded. Therefore, when the data size is large, these single-machine based tools are inefficient and time-consuming.

3 Methods

3.1 MapReduce Methodology

The methodology consists of two MapReduce phases, as shown in Fig. 1. In the first phase, parallel mappers perform tasks of loading, filtering, key-value mapping, binning and parallel sampling. Sampling is to approximate the distribution of SNP records over the whole genome, generating partition list files to divide the whole genome into splits of comparable size assigned to following tasks. In the second MapReduce phase, parallel MapReduce jobs corresponding to specified chromosomes are created. Within each job, a partitioner shuffles key-value paired records to appropriate reducers by referring to partition lists. Finally, reducers carry out sorting and merging, converting them to sorted TPED files.

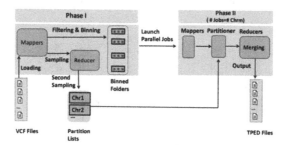

Fig. 1. The workflow chart of MapReduce methodology.

3.2 HBase Methodology

As shown in Fig. 2, The HBase methodology is divided into three phases, namely the sampling phase, the bulk loading phase, and the exporting phase. The sampling phase is similar to its counterpart in the MapReduce methodology for pre-splitting the table into regions of approximately equal sizes. During the bulk loading, key-value paired records from previous phase are sorted and saved as HBase's native storage file format, HFile, which are then readily loaded into a table. In the exporting phase, a table scan with specified chromosomes are performed on the table. This scan involves launching

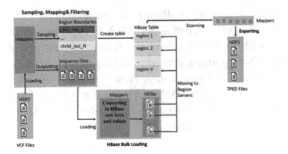

Fig. 2. The workflow chart of HBase methodology.

parallel mappers each of which retrieves records from a region, do the joining and outputting of TPED files.

3.3 Spark Methodology

Our Spark methodology has one spark job with three stages as shown in Fig. 3. The first stage involves loading raw data as RDDs, filtering, and mapping RDDs to PairRDDs with keys (chromosome-SNP loci) and values (genotype). This stage ends with an reduceByKey shuffling which groups together PairRDD records with same keys and merging individual genotypes together. In the second stage, merged PairRDD records are redistributed across executors to be sorted on keys. In the final stage, sorted PairRDDs are mapped to and save as TPED files on HDFS.

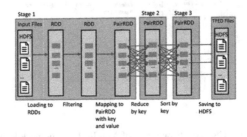

Fig. 3. The workflow chart of Spark methodology.

4   Preliminary Experimental Results

4.1   Experimental Environment

All our tests were conducted using Amazon's EMR service. Within the infrastructure, we chose EC2 working nodes of m3.xlarge type which has four High Frequency Intel Xeon E5-2670 v2 (Ivy Bridge) Processors and 15 GB memory.

4.2   Overall Performances of Cluster-Based Methodologies

As Fig. 4 shows, for all three methodologies, with fixed core number, the time cost increased almost linearly as the number of input files increased. We also noticed, with fixed input size, the more the cores, the less the time spent on merging. This was confirmed in another set of experiments in which the number of input files was kept consistently at 93 and core number was increased from 12 to 72, as shown in Fig. 5. These results have indicated a good scalability of all three methodologies on input data size and computing resources.

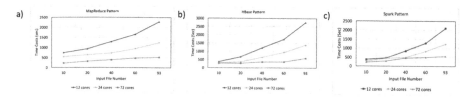

Fig. 4. The scalability of cluster-based methodologies on input size.

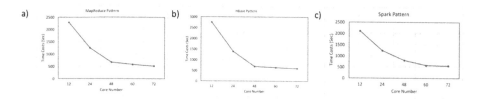

Fig. 5. The scalability of cluster-based methodologies on computing core number.

4.3   Comparisons Between Multiway-Merge and Cluster-Based Methodologies

We compared the performance in between our three cluster-based methodologies as well with the single machine multiway-merge implementation as a benchmark. Figure 6 shows their time costs on merging increasing number of files with 72 cores. As the input size increased from 10 to 93, the time cost of multiway-merge increased from 3-fold to

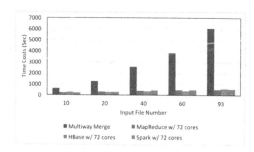

Fig. 6. Comparison of the three cluster-based methodologies with multiway-merge.

12-fold more than that of our fastest clustered-based method. As an extreme case, when merging all 690 files, the time cost of multiway merge method was 1228 min in contrast with 11.3 min of MapReduce methodology with 400 cores.

5 Discussion

While multiway-merge based software require the input data in sorted order, such a limitation can be eliminated in all three cluster-based methodologies. Moreover, the unique features the three platforms render different application scenarios. For example, MapReduce methodology is good for one time merging on large scale of data, HBase methodology fits for incremental merging, while Spark is good for carrying out subsequent statistical analysis directly on the merged results.

6 Conclusion

Our methodologies can efficiently convert large scale of VCF files into TPED format. All three Hadoop patterns show good scalability on core number and input data size. More importantly, our methodologies can be generalized as a cluster-based paradigm for key-based sorted full-out-joining problems.

Acknowledgement. This work is supported in part by NSF ACI 1443054 and NSF IIS 1350885.

References

Chang, F., et al.: Bigtable: a distributed storage system for structured data. ACM Trans. Comput. Syst. **26**(2), 4 (2008)

Danecek, P., et al.: The variant call format and VCFtools. Bioinformatics **27**(15), 2156–2158 (2011)

Dean, J., Ghemawat, S.: Mapreduce: simplified data processing on large clusters. Commun. ACM **51**(1), 107–113 (2008)

Massie, M., et al.: Adam: genomics formats and processing patterns for cloud scale computing. University of California, Berkeley Technical Report, No. UCB/EECS-2013 2013; 207 (2013)

Purcell, S., et al.: PLINK: a tool set for whole-genome association and population-based linkage analyses. Am. J. Hum. Genet. **81**(3), 559–575 (2007)

Zaharia, M., et al.: Resilient distributed datasets: a fault-tolerant abstraction for in-memory cluster computing. In: Proceedings of the 9th USENIX Conference on Networked Systems Design and Implementation, p. 2. USENIX Association (2012)

An Emerging Role for Polystores
in Precision Medicine

Edmon Begoli[1]([✉]), J. Blair Christian[1], Vijay Gadepally[2],
and Stavros Papadopoulos[3]

[1] Oak Ridge National Laboratory, 1 Bethel Valley Road, Oak Ridge, TN 37831, USA
{begolie,christianjb}@ornl.gov
[2] Massachusetts Institute of Technology (MIT),
Lincoln Laboratory, 244 Wood Street, Lexington, MA 02420, USA
vijayg@ll.mit.edu
[3] TileDB, Inc., Cambridge, MA 02139, USA
stavros@tiledb.io

Abstract. Medical data is organically heterogeneous, and it usually
varies significantly in both size and composition. Yet, this data is also
a key for the recent and promising field of precision medicine, which
focuses on identifying and tailoring appropriate medical treatments for
the needs of the individual patients, based on their specific conditions,
their medical history, lifestyle, genetic, and other individual factors. As
we, and a database community at large, recognize that a "one size does
not fit all" solution is required to work with such data, we present our
observations based on our experiences, and the applications in the field
of precision medicine. We make the case for the use of polystore architecture; how it applies for precision medicine; we discuss the reference
architecture; describe some of its critical components (array database);
and discuss the specific types of analysis that directly benefit from this
database architecture, and the ways it serves the data.

Keywords: Polystore · Precision medicine · Genomics · Array database

1 Introduction

Medicine, and the discipline of professional medical care, deals with the complex
biological, demographic, behavioral, therapeutic, procedural, and other phenom-
ena related to the health of the patients, so the data resulting from the related
processes is inherently complex. It is complex in structure, and asymmetric in
size (e.g., the size of clinical images vs. the size of electronic health records is often
out of balance by at least an order of magnitude). In a contemporary medical
system, the typical process of medical care centers around the physician who is
serially treating groups of patients based on their conditions, and using standard,
cohort-specific medical procedures and treatments. This long-standing system
works well with well-understood and relatively simple conditions. With complex
conditions, such as cancer, medical and life sciences community is increasingly

© Springer International Publishing AG 2017
E. Begoli et al. (Eds.): DMAH 2017, LNCS 10494, pp. 41–52, 2017.
DOI: 10.1007/978-3-319-67186-4_5

applying the practices of precision medicine as the more effective and precise way to treat patients.

Precision medicine [23] promotes tailoring the available therapies for the specific patients, and for the differentiating characteristics of specific conditions. To make this possible, the most complete, practically available data on patients and conditions is needed. To be effective, clinicians need to make decisions based on the clinical/medical records data, laboratory results, medical imagery data, oncology and radiology reports, clinical trials data, genomic sequences, and, increasingly, personal wearable devices data [29].

To support this approach, we need means to manage, access, and analyze all this data productively and efficiently. Despite their utility and general validity, the traditional approaches to data organization and preparation for data analysis are not adequate. We recognize that there is not a single system in existence, or even a conglomerate of systems, that would allow precision medicine by monolithically manging the data. Instead, we recognize that it is best to keep the data in the original systems that manage this data optimally, and then find ways to access it in a uniform, but federated way. The approach presented in about doing exactly that (federated data management). It is based on the implementation of the original *polystore* [13] concept in the context of precision medicine – i.e. an implementation of a modern, federated heterogeneous database system for data relevant to precision medicine research – discrete, relational; text; genomic; images; and times series data.

2   Polystore Architectures

Polystore systems refer to systems that integrate heterogenous database engines and multiple programming languages [14]. Thus, a polystore system can provide a single interface between SQL, NoSQL and NewSQL engines while supporting varying levels of location transparancy and semantic completeness. We briefly contrast polystore systems with Federated, Polyglot, and Multistore systems. Federated systems such as Garlic [8], IBM DB2 [16] and others [31] are largely designed to integrate multiple homogenous database engines via a single programming language such as SQL. Multistore system such as CloudMdSQL [19] are largely designed to integrate multiple heterogenous database engines via a single programming language. Polyglot systems are largely designed to integrate homogenous database engines via heterogenous programming or query languages. Of course, there are many existing system which may fall into one or more of these categories.

Polystore systems leverage techniques from the parallel computing [18,20] and parallel database [11] communities such as replication, partitioning and horizontally scaled hardware. Protype polystore systems such as BigDAWG, Myria [33] and the AWESOME polystore [10] have been shown to scale well with heterogenous data federated across disparate database engines. Another important component of Polystore systems is the ability to support multiple programming languages. The desire for such a feature is described in [32].

2.1   BigDAWG Architecture

The BigDAWG system is a prototype polystore system with support for SQL, NoSQL and NewSQL database management systems and has been applied to complex scientific [22] and medical data [13]. BigDAWG is open-source and available at https://bigdawg.mit.edu. Currently, BigDAWG supports a number of engines such as MySQL, PostgreSQL, SciDB, Accumulo and S-Store. A notional BigDAWG architecture applicable to a precision medicine application is shown in Fig. 1.

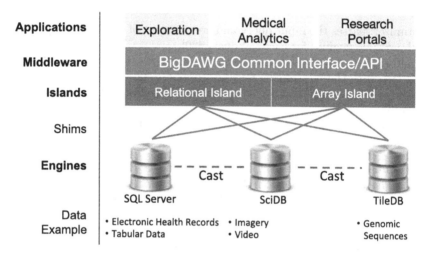

Fig. 1. Notional BigDAWG architecture for precision medicine

At the bottom of Fig. 1, BigDAWG can support a number of database engines such as SQL Server, TileDB and SciDB that are relevant to precision medicine applications (connectors to SQL Server and TileDB are being developed). These storage engines are organized into a number of islands which encapsulate a data model, a set of operations, and a set of candidate storage engines. An island provides location transparency among its associated storage engines. For example, the **array** island may convert a BigDAWG query into an *Array Query Language (AQL)* query that can be understood by SciDB, TileDB or by developing stored procedures in other systems such as SQL Server that support AQL operations.

A shim connects an island to one or more storage engines. The shim translates queries expressed in terms of the operations defined by an island into the native query language of a particular storage engine. Thus, for two arrays (or tables) A and B, the shim may convert the **AQL** query **A*B** to something like the following in SQL:

```
SELECT A.row_num, B.col_num, SUM(A.value*B.value)
FROM A,B
WHERE A.col_num = B.row_num
GROUP BY A.row_num, B.col_num;
```

In order for a query to best match the underlying storage system, we expect that a single BigDAWG query will span multiple storage engines and possibly islands. The BigDAWG common interface is designed to support such operations. This middleware supports a common application programming interface to a collection of storage engines via a set of islands. The middleware consists of a number of components:

- Optimizer: parses the input query and creates a set of viable query plan trees with possible engines for each subquery [30].
- Monitor: uses performance data from prior queries to determine the query plan tree with the best engine for each subquery [9].
- Executor: figures out how to best join the collections of objects and then executes the query [17].
- Migrator: moves data from engine to engine when the plan calls for such data motion [12].

For example, the following BigDAWG query moves data from SciDB to PostgreSQL. The bdarray() portion of the query filters all entries in the SciDB array myarray with dim1>150. The bdcast() portion of the query tells the middleware to migrate this resultant array to a table called tab6 with schema (i bigint, dim1 real, dim2 real) to a database in the relational island. The final bdrel() portion of the query selects all entries from this resultant table in PostgreSQL.

```
bdrel(
    select * from
     bdcast(
        bdarray(filter(myarray,dim1>150))
        , tab6
        ,'(i bigint, dim1 real, dim2 real)'
        , relational))
```

Further details of these components, usage, and example queries can be found in [14,15].

2.2 BigDAWG Performance

BigDAWG has been applied to a number of problems in the scientific [22] and medical [13] domains. In our experiments, we have demonstrated that using a *many sizes* solution can greatly improve the performance of an overall analytic system. For example, in [14], we describe an analytic performed on ECG signals from the MIMIC II [28] dataset. In [22], we demonstrate how BigDAWG can be used to help researchers correlate genetic sequences and metadata.

2.3   BigDAWG for Precision Medicine

Precision medicine applications are a natural candidate for polystore solutions. Healthcare applications at organizations such as the Department of Veterans Affairs rely on data from a multitude of data sources persisted on disparate database management systems. For example, medical electronic health records (EHR) are stored in a clinical data warehouse that may leverage a technology such as Microsoft SQL Server. Images derived from sources such as X-Ray or magnetic resonance imaging (MRI) are stored in image specific databases. Other application specific systems such as research databases and genetic databases increase the complexity of developing applications that touch the various data components. For example, trying to analyze the fitness activities associated with particular patients which exhibit certain genetic anomalies may require aggregating data from research database(s), genetic databases, and the clinical data warehouse. Current practices rely on manual data transformations and data modeling which can take significant developer effort and lead to degraded overall performance. Polystore systems such as BigDAWG on the other hand can allow developers to keep data in the database that they are best suited for. For example, EHR records can sit in a federated systems such as SQL Server, genetic sequences in a system such as SciDB or TileDB and streaming data in a system such as S-Store.

2.4   Array Databases – A Critical Component of a Polystore for Precision Medicine Research

Array databases are optimized to store and process data that are naturally represented as objects in a multi-dimensional space. An array has *dimensions* and *attributes*. Each dimension has a domain, and all dimension domains collectively orient the logical space of the array. Each array element, called a *cell*, is uniquely identified by a combination of dimension domain values, referred to as *coordinates*. A cell may either be empty (null) or contain a tuple of attribute values. Each attribute is either a primitive type (e.g., `int`), or a fixed or variable-sized vector of a primitive type. Arrays can be *dense*, where every cell contains an attribute tuple, or *sparse*, where the majority of the cells are empty.

Figure 2 shows two example arrays, one dense one sparse, with two dimensions (`rows` and `columns`). Both dimensions have domain [1,4]. The arrays have two attributes, `a1` of type `int32` and `a2` of type variable `char` (i.e., `string`). Each cell is identified by its coordinates, i.e., a row index and a column index. Empty cells are shown in white and nonempty cells in gray. In the dense array, every cell has an attribute tuple, e.g., cell (4,4) contains 15 and `pppp`, whereas several cells in the sparse array are empty, e.g., cell (4,1).

RasDaMan [5] was among the first array databases. It stores array data as binary large objects (BLOBs) in PostgreSQL [3], and enables query processing via a SQL-based array query language called RasQL. Array storage and access has to go through PostgreSQL, which is typically quite slow. Moreover, RasDaMan is designed mainly for dense arrays. SciDB [7] is a "pure" array

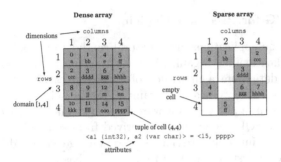

Fig. 2. The array model

database which implements its own storage and processing engine and provides both dense and sparse array support. TileDB [26] is an emerging array system. Contrary to SciDB that is a monolithic, integrated database software, TileDB is implemented as a lightweight array storage C library that offers excellent, native performance and binds with the various programming interfaces like Python and R, but without incurring the extra overheads for establishing database connections (ODBC/JDBC), or parsing a query with some additional array languages (AQL, etc.). (Results presented in [26] demonstrate the performance superiority of TileDB for both dense and sparse array storage as compared to another well-known array (SciDB), or other commercial databases.)

TileDB achieves high performance in two ways. First, it provides a novel format for storing dense and sparse arrays, while providing the user a unified C API interface for querying an array. Coupled with efficient multi-dimensional index structures, this format offers rapid ingestion and searching capabilities. Second, similar to Vertica [21] and SciDB, it follows a *columnar* physical layout for its attributes, i.e., it splits each cell tuple into its separate attributes, storing contiguously the values of each attribute in a separate file on the disk. This enables fast sub-selecting over the attributes and offers optimization opportunities in compression, memory management, and execution (e.g., through SIMD vectorization when processing contiguous attribute values).

TileDB has been already used successfully in the genetics domain. In a collaboration between Intel Health and Life Sciences and the Broad Institute, it has been demonstrated [2] that TileDB is ideal for storing and processing genetics data that come in the so called Variant Call Format (VCF). This effort has led to a software built on top of TileDB, called GenomicsDB [1], which is now part of the newest genome analysis toolkit of the Broad Institute, called GATK 4.0 [4]. Briefly stated, the VCF format captures structural variations of a certain individual's genome with respect to a reference genome. The human genome can be practically perceived as a long string consisting of approximately 3 billion letters, where each letter stands for nucleotide A, T, C, or G. An individual's VCF data file can then be perceived as a super sparse vector (of approximately 0.1% density) where the domain is 3 billion elements and a nonempty cell corresponds to some structural variation carrying a multi-attribute tuple. Similarly, the VCF files of an entire population can be collectively perceived as a 2D array with 3 billion columns and as many rows as the number of individuals in the population. It is the extreme sparsity that TileDB was able to take advantage when ingesting VCF file collections, in addition to the compressibility of its columnar format, that resulted in extreme performance gains versus previous approaches.

TileDB is a generic array management system and, as such, it can be used in other use cases where the data are multi-dimensional. One of the domains we are currently exploring is storage and processing of medical imaging (e.g., CT and MRI scans). Such data are naturally modeled as 2D or 3D dense or sparse, with each cell representing a pixel in an image snapshot.

3 A Statistical Viewpoint - The Importance of Uniform Access to Data for Reproducible, Automated Analyses in Precision Medicine

The polystore represents a natural progression in the development of analytics architectures not just in healthcare, but across many areas with complex, heterogeneous data. In healthcare, as practitioners demand more analyses to determine the best care for their patients, standards and automation in data preparation and analysis are needed. The polystore architecture fills this emerging gap and provides a needed foundation for repeatable, reproducible and automated analyses not only in support of precision medicine, but also in support of other types of healthcare improvements. There are two main phases to precision medicine, (1) building statistical models using historical data, and (2) applying these models to new patients in real time. Historically, these two steps have been independent. However, the polystore and new statistical methodology for adaptive design [25, 35] and pragmatic trials [27] enables these two steps to be intertwined at scale and for the first time, so that optimal treatment models can be updated as new patients enter the system and outcomes can be monitored in real time to detect unusual variation [6].

3.1 Key Statistical and Analytic Problems Driving the Data Architecture

There are several key features for precision medicine and care improvement that are common to all endeavors. Across methods such as genome wide association studies (GWAS), optimal treatment of acute conditions, optimal treatment of chronic conditions, and surveillance of populations to evaluate outcomes, data must be fitted into a mathematical matrix for input into the analytic software, which is based on numerical linear algebra. The basis of precision medicine is understanding what genetic and environmental factors drive health outcomes. To convert these heterogeneous genetic and environmental variables into a mathematical matrix, a very large number of sophisticated tools are needed to take raw, silo-ed data, and create mathematical representations called features (machine learning) or covariates (statistics). Further, this process needs to be automated, reproducible, and repeatable, capable of answering not just these questions, but also supporting many ancillary requests as well [6,24]. We recognize that the polystore architecture helps overcome these many issues with the *status quo* approach.

In most cases, both in the governmental computing, and in the healthcare industry, data is collected from many heterogeneous sources including traditional relational databases but also including images, text, genomic, and spatial sources, and tools from computer vision, bioinformatics, the census, NLP domains must be used to create covariates/features which are the inputs to learning algorithms. Not only do these feature representations need to be made, but they also need to have the correct linkages so that they can be joined appropriately. For example, socioeconomic data from the US Census or American Community Survey may be available at spatial resolutions such as tract, block group or block, while health survey data may be available at the county or zip code level. In this case, a spatial disaggregation or change of support may be needed. Similarly, the resolution of some genomic data may be relatively coarse, and it may be necessary to use population data on linkage disequilibrium to estimate the haplotypes or alleles of interest in between observed genomic data. The polystore approach helps simplify, integrate, and automate the solutions to these problems by offering a common query langue model which can be further used by the common analytics languages (R, SAS, etc.)

3.2   Statistical Foundations of Precision Medicine and Healthcare Improvement

There are several classes of problems behind precision medicine and healthcare improvement in general. The foundation of the GWAS methodology is the assumption that all variation in phenotypes can be accounted for by genetic, environmental, and epigenetic factors. However, the data available is almost never at the appropriate level of genomic or environmental granularity to begin an analysis. Historically, this meant that research involved creating very specific transformations to answer a limited number of questions and avoided any overhead associated with creating a data store optimized for answering a large number of questions.

Similarly, determining the optimal treatment regimes for acute and chronic conditions requires that temporal representations of patients be created using common data models. More specifically, the competing factors of 'big n' and specificity of condition come into play that demand a hierarchical common data model so that the strongest results can be made at the appropriate level of the hierarchy. In other words, most practitioners want clinical decision support results very specific to their patient, "What treatment regime will lead to the best outcome for treating a specific type of lymphoma in an X year old, with Y comorbidities, and is positive/negative for biomarkers Z?" However, there may be no patients in the historical population with comparable age, gender, comorbidities, and biomarker combinations, so the physician needs to traverse the data models to find a sample size large enough to detect a specific effect size.

When these pieces of data are located in different databases and must be evaluated at different levels of the data model, a polystore may provide the only practical approach.

3.3 Desired Architecture for Healthcare Analysis and Improvement

Currently, a physician may do a literature search to try to find an optimal treatment, or may use an established clinical pathway that does not take into account important data, such as patient biomarkers. Our vision is to create an architecture capable of automating these processes by creating an advanced analytics infrastructure, as opposed to merely supporting one off investigations. This would enable not only the automation of optimal treatment modeling, but also the capability to pre-cache all possible optimal treatment models for a given class of outcomes, treatments, covariates and sample size/effect size. Further, it enables adaptive experiments to be conducted in real time. Combining data sources is but the first of many steps in this endeavor. There are many key features and services that must be present to achieve our goal. Specifically, we must be able to combine (e.g., link) data sources together efficiently at the appropriate granularity (e.g., hierarchical common data models). This process requires large amounts of metadata and functions to map data to common data model entities.

3.4 Polystore Capabilities for Precision Medicine

The polystore offers automation, repeatability, and reproducibility advantages compared to the *status quo* for precision medicine. The key feature of a polystore, such as BigDAWG and its instantiations (e.g. VA Polystore) is the universal interface to a federated collection of data. This feature enables a standard, so that the creation of features is concise, repeatable, and reproducible. This allows models and analyses to be run with minimal overhead. In fact, these analyses can be automated, pre-cached, and can even enable adaptive statistical experimental designs, all without replicating the data. This is a major breakthrough for precision medicine applications.

4 Emerging Approaches and Ongoing Research

The approach and the case we presented in this paper was focused on advanced data management methods, and the motivating statistical and machine learning applications. We observe an emerging body of work that takes this a step further, towards integrated data management and machine learning and artificial intelligence techniques where the federated data management of a polystore is going to be integrated with the machine learning functions. This is the work that we are currently pursuing at our research institutions.

Furthermore, there are ongoing developments with the database technologies that introduce hardware acceleration techniques [34] that leverage the same underlying hardware subsystems as the machine learning libraries. We expect to see a convergence of these two technologies into a database solution for healthcare that will incorporate, for example, both GPU-accelerated machine learning functions, and database routines. To this end, we expect to see the incorporation of some kind of machine learning functions into the Polystore API and its syntax.

There is also an ongoing effort to define the unifying, theoretical algebra for Polystores which would make the implementation of any number of sources theoretically well-founded, including the development of algebra for text-searching queries.

5   Future Work

Both precision medicine and polystores are relatively new concepts in their own domains. Consequently, there is a rich ongoing and future work expected to happen, and to promote the advancements in both areas. In terms of our future work on polystores for precision medicine, we will continue to improve the performance and robustness of the polystore system, specially for the scientific data; we will grow the support for diverse Polystore *islands*, and we will continue to make improvements to the usability of the system. In terms of specific technical advances, we will accelerate work towards the integration of the textual and imagery medical data; we will optimize shimming approaches for non-relational data, and we will make contributions to the work on the unifying, polystore algebra.

Acknowledgments. This manuscript has been in part authored by UT-Battelle, LLC, under contract DE-AC05-00OR22725 with the U.S. Department of Energy, and under a joint program (MVP CHAMPION), between the U.S. Department of Energy, and the U.S. Department of Veterans Affairs.

The authors would like to thank the Intel Science and Technology Center (ISTC) for Big Data and the BigDAWG contributors (https://bigdawg.mit.edu/contributors) for their role in developing the BigDAWG system.

References

1. GenomicsDB. https://github.com/Intel-HLS/GenomicsDB
2. Intel-Broad Collaboration. http://genomicinfo.broadinstitute.org/acton/media/13431/broad-intel-collaboration
3. PostgreSQL. http://www.postgresql.org
4. Unboxing GATK4. https://gatkforums.broadinstitute.org/gatk/discussion/9644/unboxing-gatk4
5. Baumann, P., Dehmel, A., Furtado, P., Ritsch, R., Widmann, N.: The multidimensional database system RasDaMan. In: SIGMOD (1998)
6. Benneyan, J.C., Lloyd, R.C., Plsek, P.E.: Statistical process control as a tool for research and healthcare improvement. Qual. Saf. Health Care **12**(6), 458–464 (2003)

7. Brown, P.G.: Overview of SciDB: large scale array storage, processing and analysis. In: SIGMOD (2010)

8. Carey, M.J., Haas, L.M., Schwarz, P.M., Arya, M., Cody, W.E., Fagin, R., Flickner, M., Luniewski, A.W., Niblack, W., Petkovic, D., et al.: Towards heterogeneous multimedia information systems: the Garlic approach. In: Proceedings of the Fifth International Workshop on Research Issues in Data Engineering, 1995: Distributed Object Management. RIDE-DOM 1995, pp. 124–131. IEEE (1995)

9. Chen, P., Gadepally, V., Stonebraker, M.: The bigdawg monitoring framework. In: High Performance Extreme Computing Conference (HPEC), 2016 IEEE, pp. 1–6. IEEE (2016)

10. Dasgupta, S., Coakley, K., Gupta, A.: Analytics-driven data ingestion and derivation in the AWESOME polystore. In: 2016 IEEE International Conference on Big Data (Big Data), pp. 2555–2564. IEEE (2016)

11. DeWitt, D., Gray, J.: Parallel database systems: the future of high performance database systems. Commun. ACM **35**(6), 85–98 (1992)

12. Dziedzic, A., Elmore, A.J., Stonebraker, M.: Data transformation and migration in polystores. In: 2016 IEEE High Performance Extreme Computing Conference (HPEC), pp. 1–6. IEEE (2016)

13. Elmore, A., Duggan, J., Stonebraker, M., Balazinska, M., Cetintemel, U., Gadepally, V., Heer, J., Howe, B., Kepner, J., Kraska, T., et al.: A demonstration of the BigDAWG polystore system. Proc. VLDB Endow. **8**(12), 1908–1911 (2015)

14. Gadepally, V., Chen, P., Duggan, J., Elmore, A., Haynes, B., Kepner, J., Madden, S., Mattson, T., Stonebraker, M.: The BigDAWG polystore system and architecture. In: 2016 IEEE High Performance Extreme Computing Conference (HPEC), pp. 1–6. IEEE (2016)

15. Gadepally, V., OBrien, K., Dziedzic, A., Elmore, A., Kepner, J., Madden, S., Mattson, T., Rogers, J., She, Z., Stonebraker, M.: Version 0.1 of the BigDAWG Polystore System. arXiv preprint arXiv:1707.00721 (2017)

16. Gassner, P., Lohman, G.M., Schiefer, K.B., Wang, Y.: Query optimization in the IBM DB2 family. IEEE Data Eng. Bull. **16**(4), 4–18 (1993)

17. Gupta, A.M., Gadepally, V., Stonebraker, M.: Cross-engine query execution in federated database systems. In: 2016 IEEE High Performance Extreme Computing Conference (HPEC), pp. 1–6. IEEE (2016)

18. Hudak, D.E., Ludban, N., Krishnamurthy, A., Gadepally, V., Samsi, S., Nehrbass, J.: A computational science IDE for HPC systems: design and applications. Int. J. Parallel Prog. **37**(1), 91–105 (2009)

19. Kolev, B., Bondiombouy, C., Valduriez, P., Jiménez-Peris, R., Pau, R., Pereira, J.: The cloudmdsql multistore system. In: Proceedings of the 2016 International Conference on Management of Data, pp. 2113–2116. ACM (2016)

20. Krishnamurthy, A., Samsi, S., Gadepally, V.: Parallel MATLAB techniques. In: Image Processing. InTech (2009)

21. Lamb, A., Fuller, M., Varadarajan, R., Tran, N., Vandiver, B., Doshi, L., Bear, C.: The vertica analytic database: C-store 7 years later. Proc. VLDB Endow. **5**(12), 1790–1801 (2012)

22. Mattson, T., Gadepally, V., She, Z., Dziedzic, A., Parkhurst, J.: Demonstrating the BigDAWG polystore system for ocean metagenomics analysis. In: CIDR (2017)

23. Mirnezami, R., Nicholson, J., Darzi, A.: Preparing for precision medicine. N. Engl. J. Med. **366**(6), 489–491 (2012)

24. Ng, K., Ghoting, A., Steinhubl, S.R., Stewart, W.F., Malin, B., Sun, J.: PARAMO: a PARAllel predictive MOdeling platform for healthcare analytic research using electronic health records. J. Biomed. Inform. **48**, 160–170 (2014)
25. Palmer, C.R.: Ethics, data-dependent designs, and the strategy of clinical trials: time to start learning-as-we-go? Stat. Methods Med. Res. **11**(5), 381–402 (2002)
26. Papadopoulos, S., Datta, K., Madden, S., Mattson, T.: The tiledb array data storage manager. Proc. VLDB Endow. **10**(4), 349–360 (2016)
27. Roland, M., Torgerson, D.J.: Understanding controlled trials: what are pragmatic trials? BMJ: Br. Med. J. **316**(7127), 285 (1998)
28. Saeed, M., Villarroel, M., Reisner, A.T., Clifford, G., Lehman, L.-W., Moody, G., Heldt, T., Kyaw, T.H., Moody, B., Mark, R.G.: Multiparameter intelligent monitoring in intensive care II (MIMIC-II): a public-access intensive care unit database. Crit. Care Med. **39**(5), 952 (2011)
29. Safran, C., Bloomrosen, M., Hammond, W.E., Labkoff, S., Markel-Fox, S., Tang, P.C., Detmer, D.E.: Toward a national framework for the secondary use of health data: an American Medical Informatics Association White Paper. J. Am. Med. Inform. Assoc. **14**(1), 1–9 (2007)
30. She, Z., Ravishankar, S., Duggan, J.: Bigdawg polystore query optimization through semantic equivalences. In: 2016 IEEE High Performance Extreme Computing Conference (HPEC), pp. 1–6. IEEE (2016)
31. Sheth, A.P., Larson, J.A.: Federated database systems for managing distributed, heterogeneous, and autonomous databases. ACM Comput. Surv. (CSUR) **22**(3), 183–236 (1990)
32. Stonebraker, M., Cetintemel, U.: "one size fits all": an idea whose time has come and gone. In: Proceedings of the 21st International Conference on Data Engineering. ICDE 2005, pp. 2–11. IEEE (2005)
33. Wang, J., Baker, T., Balazinska, M., Halperin, D., Haynes, B., Howe, B., Hutchison, D., Jain, S., Maas, R., Mehta, P., et al.: The myria big data management and analytics system and cloud services. In: CIDR (2017)
34. Yong, K.K., Karuppiah, E.K., See, S.C.-W.: Galactica: a GPU parallelized database accelerator. In: Proceedings of the 2014 International Conference on Big Data Science and Computing, p. 10. ACM (2014)
35. Zhou, X., Liu, S., Kim, E.S., Herbst, R.S., Lee, J.J.: Bayesian adaptive design for targeted therapy development in lung cancera step toward personalized medicine. Clin. Trials **5**(3), 181–193 (2008)

Online Mining of Health Related Data

Social Media Mining to Understand Public Mental Health

Andrew Toulis and Lukasz Golab[✉]

University of Waterloo, Waterloo, ON N2L 3G1, Canada
{aptoulis,lgolab}@uwaterloo.ca

Abstract. In this paper, we apply text mining and topic modelling to understand public mental health. We focus on identifying common mental health topics across two anonymous social media platforms: Reddit and a mobile journalling/mood-tracking app. Furthermore, we analyze journals from the app to uncover relationships between topics, journal visibility (private vs. visible to other users of the app), and user-labelled sentiment. Our main findings are that (1) anxiety and depression are shared on both platforms; (2) users of the journalling app keep routine topics such as eating private, and these topics rarely appear on Reddit; and (3) sleep was a critical theme on the journalling app and had an unexpectedly negative sentiment.

Keywords: Text mining · Social media mining · Public mental health

1 Introduction

Many applications of social media involve text mining, such as understanding user interests, customer reviews, and sentiment around news events. We discuss an application of social media text mining in the context of understanding public mental health. This is an increasingly important application domain: prevalence of mental health conditions is increasing, and so is the amount of data we have to understand these conditions [2]. While text data has been analyzed in great depth for marketing purposes, there remains a large opportunity in using text data to understand public mental health.

Several researchers have identified social media data, such as Twitter posts, as a valuable source for mental health signals [3,6]. However, there remain critical gaps in our ability to understand mental health. People may not be willing to share mental health content publicly, especially on the largest social media platforms, which are associated with a personal identity.

In this paper, we perform a mental health analysis on stigmatized topics by taking advantage of a unique dataset from a social media app for posting journal entires and sharing and tracking moods (referred to as the journalling app). We discover issues that are not widely discussed on other social media such as sleep. In addition, the journalling app requires users to track their moods in each

© Springer International Publishing AG 2017
E. Begoli et al. (Eds.): DMAH 2017, LNCS 10494, pp. 55–70, 2017.
DOI: 10.1007/978-3-319-67186-4_6

journal. Hence, we are equipped with user-labeled sentiment, which otherwise is difficult to estimate.

By comparing the journalling app dataset to other social media (Reddit), we identify unique discussions that the mood-tracking community attracts. Furthermore, the journalling dataset is split into two segments: users may share their journals publicly or keep them private. Hence, we are able to understand which mental health issues are shared more publicly than others.

The questions we seek to answer include:

1. Does the journalling dataset cover a different set of topics than those discussed on other social media such as Reddit?
2. Are some topics shared more publicly than others? Are some topics kept private?
3. How is sentiment related to topics? Which topics elicit sad or happy feelings?

To answer these questions, we use a text mining methodology to derive topics from journals. Furthermore, we take advantage of already labelled moods to perform sentiment analysis. To summarize, we make the following contributions:

1. We apply text mining to a unique dataset of journals and associated moods that has not been studied before. We discover a set of mental health topics that are not frequent on other social media.
2. We quantify which topics are more public than others to identify gaps in available social media data for analyzing public mental health.
3. We compare how user-labeled moods vary across topics to identify important aspects of mental health that require attention.

The remainder of this paper is organized as follows. We discuss context and related work in Sect. 2; we describe our datasets in Sect. 3; we explain our methodology in Sect. 4 followed by our results in Sect. 5; and we conclude in Sect. 6.

2   Context and Related Work

This paper is related to two bodies of work: social media text mining and studies of public mental health. In the context of text mining, there are standard analysis techniques that enable topic modelling and sentiment analysis of social media posts. In the mental health domain, these techniques have had several successes including detecting users expressing suicidal thoughts on social media [11]. We also use standard topic modelling techniques but we utilize them to perform novel topic comparisons between datasets.

Mental health studies traditionally collect information via health care professionals, which is a costly process and only allows for analysis of a small subset of the public. A significant opportunity for understanding mental health through social media data has been identified by Harman et al. [6]. They focused on specific mental health conditions, and despite low incident rates they found a wealth of data on social media. They concluded that individual and population-level mental health analysis can be made significantly cheaper and more efficient than current methods.

Traditional therapy and studies of mental health conditions heavily utilize linguistic signals. In Diederich et al. [5], text processing is used to detect mental health conditions such as schizophrenia by analyzing conversations between patients and their psychiatrists using clustering algorithms and sentiment classifiers. The drawback of most studies utilizing doctor-patient data is privacy concerns and smaller datasets. These studies tend to be more ad-hoc due to the size of data. Another large opportunity for mental health data analysis is in electronic medical records. For example, natural language processing (NLP) was used to improve classification accuracy of depression in mood states of patients based on medical records [13].

The creation of a social media corpus for mental health data could significantly improve mental health research [2]. There are two main methodologies for analyzing mental health signals in social media using linguistic signals. The first relies on hand-crafted lexicons containing connotations and strengths of words. For example, the Linguistic Inquiry Word Count (LIWC) lexicon has been used to help clinicians understand mental states given a patient's writings [6]. The disadvantage of this method is that lexicons like LIWC cover a very small portion of possible language used in informal contexts such as social media.

The second common method is to train a language classifier model. This technique is limited when ground-truth labels are not available. Existing work has attempted to approximate labels, and a conservative labeling approach is to filter for users who self-identify with a condition. In particular, previous work searched for statements such as "I was diagnosed with X" [2]. However, there are caveats that the authors identify with this approach. In particular, only a small sample of people would publicly self-identify with a mental health condition. Despite this, through a language model they were able to compare language uses across specific mental health conditions [2].

Alternative pipelines for acquiring labels to model social media text include crowd-sourcing and developing custom apps [2]. Crowd-sourcing involves surveying users. In a previous study, surveys were used to study mental health trends in undergraduate students [4]. In our study, we also identify school-related issues (among other things) as a frequent topic discussed by journallers. While successes with surveys have been made, having users agree to honestly share their personal information is difficult and it can be costly to solicit other data such as social media from surveyed users.

On the other hand, apps that interact with social media such as Facebook can be used to collect personality information and grant access to public status updates. However, signals that are important for mental health analysis are not typically shared on Facebook.

While existing work has focused on traditional social media, talking about difficult issues is not common on these platforms. On the other hand, the dataset we are studying is specifically designed for mood tracking. The journalling app's goal is to de-stigmatize the expression of mental health. It is fully anonymous, and hence includes topics that are typically considered taboo on personally identifiable social media platforms. Moreover, the dataset is a combination of both public and private journals, allowing for more private topics to be mentioned frequently.

Furthermore, instead of focusing on specific mental health conditions, we choose to take a broader look into the state public mental health. We demonstrate that simple, interpretable signals can be derived from our dataset. Furthermore, we use sentiment labeled by users to avoid relying on custom lexicons.

In particular, one of our most important findings is a large issue with sleep. In a study that correlated sleep problems with mental health problems, it was found that patients are much more able to identify when they have an issue with their sleep and more willing to reveal it to their doctors than a potential mental health concern [9]. Furthermore, patient self-perception of sleep issues was strongly associated with health issues, which demonstrates that people are able to accurately identify when a real problem is present. While wide-scale studies of sleep data using social media have not been performed, there is an increasing prevalence of sleep-tracking mobile apps and tools for analyzing the quality of an individual's sleep [7].

While traditional social media has helped people connect with friends and family, anonymous social media services are becoming increasingly used by people for sharing personal stories and looking for advice [14]. These communities are growing as the general public becomes comfortable sharing more information online, and benefit people who are unable or do not want to see a doctor in person to talk about mental health [14]. We believe that these types of datasets will become increasingly important to analyze for researchers. As such, we explore another anonymous social media platform, Reddit, and compare mental health topics discussed on Reddit to those written about on the journalling app.

3 Data

We analyze two datasets: (1) user communities on Reddit and (2) journals from a mental health journalling mobile app. We omit the name of the app for privacy, and we refer to it as the "journalling app".

Reddit is a social media platform that was originally used for sharing and rating content such as news, documentaries and music. Users post in and subscribe to self-organized communities known as subreddits; subscribing to a subreddit allows a user to view all posts from that subreddit. An advantage of analyzing Reddit data is that the subreddits are labelled according to their topics. Utilizing curated lists from volunteer Reddit users, we crawled all subreddits related to mental health, as well as all subreddits linked by these communities.

The second dataset consists of anonymized journal posts from a mobile app designed to help people track their moods and share them anonymously if they desire. For each journal post, the app requires the user to label the journal post with at least one mood selected from a pre-populated list including "happy",

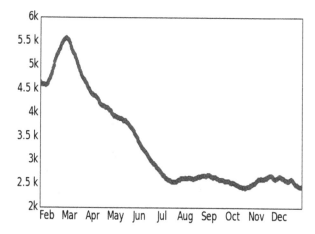

Fig. 1. Number of journals posted over time.

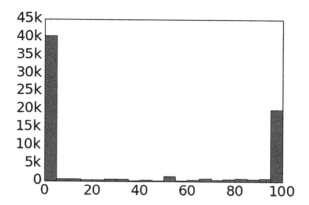

Fig. 2. Number of users plotted against the percentage of journals posted publicly.

"sad", etc. We obtained all journals, and the associated moods, written between January 2016 and January 2017. This amounts to over 1.2 million journals written by approximately 75,000 users.

Figure 1 plots the number of journals posted over time. Most of the journals were written in the first half of 2016, although we inspected topic distributions per month and did not find seasonal effects. Towards the beginning of 2016, many new users registered on the app and eventually stopped using it. Like weight-loss and productivity apps, we believe this influx is tied to users looking to improve their habits as a New Year's resolution.

Each journal can be set to be private or public (visible to all other users of the app). Roughly one third of all journals are public. Figure 2 plots the number of users on the y-axis versus the percentage of journals they posted publicly. Most users are either mostly private or mostly public.

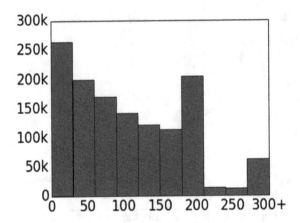

Fig. 3. Distribution of length of journals, with zero length journals included.

Most journals are relatively short, just like Twitter posts that are at most 140 characters. The average length of a journal with text in it is 128 characters; there are roughly 100,000 journal that have no text, only a mood label. We observed that private users tend to write journals that are slightly, but statistically significantly, longer than those written by public users by approximately 10 characters. Figure 3 shows the distribution of journal lengths, where the spikes correspond to 0 length (mood only), 200 characters (the default limit set by the app) and 300 characters (set as the maximum for visualization purposes).

Users of the app can optionally enter their location, age and gender. While most users did not enter this information, we found that those who revealed their location are mostly from North America, those who revealed their gender are predominantly female, and those who revealed their age have an average age of 25.

4   Methodology

The goal of this analysis is to understand public mental health by mining social media. We want to identify common topics discussed publicly (Reddit plus public journals from the journalling app) and privately (private journals). For the Reddit dataset, we simply count the number of subscribers in each subreddit related to mental health to discover popular topics and issues. Recall that each subreddit is labelled with its topic, so topic modelling is not necessary. On the other hand, for the journalling app, each journal post is labelled with a mood but not with a topic. Below, we describe our methodology for assigning topics to journals.

ALGORITHM 1. Text mining procedure to label journals with topics.

Data: Term frequency vector for each journal computed using TF-IDF
Result: A labeled set of journals

1. Perform clustering, iterating over different numbers of clusters
2. Manually label topics and evaluate them, selecting the best performing number of clusters
3. Determine a minimum threshold for the relevance of each topic in order to prune weak topic associations
4. Select the top two most relevant topic labels per journal, if any

---

4.1 Topic Modelling

First, we removed journals with no text and those with fewer than 20 characters[1], leaving 1.1 million journals for topic modelling.

Next, we pre-processed the text using the Stanford Tweet Tokenizer, which is a "Twitter-aware" tokenizer designed to handle short, informal text [1]. We used the option that truncates characters repeating 3 or more times, converting phrases such as "I'm sooooo happyy" to "I'm soo happyy". On average, the number of tokens per journal was 27.7.

Since we are interested in topics, we removed stopwords and tokens with fewer than two letters, and we only retained nouns which appear in the WordNet corpus [10]. After this filtering, the average number of nouns per journal was 7.

Examples of frequently appearing nouns, in alphabetical order, include "anxiety", "class", "dinner", "family", "god", "job", "lunch", "miss", "school", "sick", "sleep", and "work". We then iteratively clustered the journals into topics (details below) and removed nouns that do not refer to topics such as numbers, timings (e.g., "today", "yesterday"), general feelings (e.g., "feel", "like"), proper nouns, and nouns that have ambiguous meanings (e.g., "overall", "true"). Lastly, we only retained nouns that appeared more than ten times in the dataset. This process resulted in a vocabulary of 8386 words for topic modelling. Each journal is represented as a 8386-dimensional term frequency vector, with each component denoting the term-frequency/ inverse-document-frequency (TF-IDF) of the corresponding term.

Algorithm 1 summarizes our topic modelling methodology. Given a TF-IDF term frequency vector for each journal, we run non-negative matrix factorization (NMF) [8], implemented in Python's scikit-learn package [12]. The objective of NMF is to find two matrices whose product approximates the original matrix. In our case, one matrix is the weighted set of topics in each journal, and the other is the weighted set of words that belong to each topic. Hence, each journal is represented as a combination of topics which are themselves composed of a weighted combination of words.

---

[1] We manually inspected a sample of short public journals and found that those under 20 characters long typically re-stated the mood of the user and did not refer to any specific topic.

Table 1. Final list of journal topics, with the top 6 words per topic shown. We manually assigned the topic names based on the top words.

Topic						
Work	work	focus	money	meeting	friday	shift
Love	love	heart	man	fall	world	matter
School	school	high	break	summer	test	boring
Sleep	sleep	wake	sleeping	headache	nap	waking
Sickness	sick	stomach	cold	headache	throat	gross
Missing Someone	miss	old	baby	text	heart	times
Family	family	christmas	spending	health	food	husband
Career & Finances	job	interview	money	call	move	offer
Dinner	dinner	ate	movie	evening	shopping	walk
Physical Pain	pain	headache	period	empty	body	stomach
Homework	homework	finish	test	due	break	room
Anxiety/Depression	anxiety	depression	attack	high	stress	panic
School (Activities)	class	test	yoga	dance	english	teacher
Meals	lunch	ate	food	eating	break	breakfast

We chose NMF because its non-negativity constraint aids with interpretability. In the context of analyzing word frequencies, negative presence of a word would not be interpretable. This is because we only track word occurrences and not semantics or syntax. Unlike other matrix factorization methods, NMF reconstructs each document from a sum of positive parts, which enables us to easily manually label the discovered topics.

Iterating from 4 to 40 topics, we derived 37 different topic matrices (steps 1 and 2 of Algorithm 1). Each matrix consists of one topic per row. Each topic has a positive weight for each word in the vocabulary. Stronger weights indicate higher relevance to the topic. The final topic matrix we used has 14 topics and is shown in Table 1. We show the first six words in this table for simplicity, where we sorted the words associated with each topic from highest relevance to lowest.

When judging the topic matrices, we considered the top twenty most important words per topic. Using this information, we manually labeled each row in the matrix with a corresponding topic. Furthermore, we manually evaluated each matrix based on the distinctness between topics, consistency within topics, and interpretability. During this process, we compiled a custom list of removed words that we mentioned earlier in this section. The groups of words we removed appeared as stand-alone topics that did not offer information about what the journal was about. For example, proper nouns appeared as a stand-alone topic. Other words, which we deemed too general or ambiguous, appeared across several topics and hence did not provide discriminative information.

We note that by default NMF does not enforce words to be assigned a non-zero weight to only a single topic. Using our pruning procedure, we ensured words that appeared across too many topics were removed. We did permit words with multiple meanings (for example, "high") and words that apply in different settings (for example yoga "class" versus academic "class"). We note that the most important words (based on weights) for each topic generally did not overlap, with "ate" being the exception. Our validation procedure, outlined in Sect. 4.2, ensured that the two topics "Dinner" and "Meals" were indeed distinct despite both assigning high weights to "ate".

We tested different levels of regularization to enforce sparseness in our models (see [8] for a discussion), but did not find significant differences. However, one important modification we made to regularize each topic was to make their first words only as strong as their second ones (by default, first words are stronger than second words, which are stronger than third words, and so on). This is since the most relevant word for each topic tended to be too strong of a signal, regardless of how we changed the number of topics, pre-processing procedure, or regularization in the objective function. For example, the word "love" in a journal about sports would be so strong that the journal would be labeled as relating to romantic love. Lowering the importance of first words was sufficient to eliminate the false positives we identified.

Given the final topic matrix (summarized in Table 1), the next step is to use it to assign labels to journals (steps 3 and 4 of Algorithm 1). We plotted the distribution of how important each topic was to all journals in the dataset, with importance ranging from zero to one. Each distribution had a similar shape with a clear inflection point between 0.05 to 0.15 importance. Figure 4 shows an example importance distribution for the topic "Work", where the inflection point occurs at 0.1 importance.

We used these inflection points to set minimum thresholds of importance for each topic. We ignored any topic assignments below the thresholds. Then, we obtained the top two topics per journal, if any. We chose a maximum of two topics per journal due to the generally short length of journals.

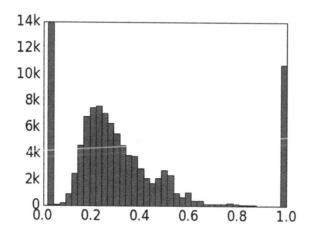

Fig. 4. Distribution of the importance of the topic "Work" to journals, with an inflection point at 10%. Zero importance has been omitted for clarity.

Using this topic modelling procedure, we assigned at least one topic to 430,000 (35% of all available) journals. The number of journals with one topic was 334,000, while 96,000 had two topics (the maximum).

4.2 Validation

To evaluate the effectiveness of our topic modelling methodology, we selected random subsets of 60 public journals for each topic and 100 public journals with no assigned topic. We manually labeled the sampled journals, looking for one of the 14 available topics, no topic or "other" topic. Including "other" allowed us to validate whether our list of manually labeled topic names were accurate and complete. We then compared our labels with those assigned by the model. For journals which were assigned two topics by the model, we considered the model correct if either one of the assigned topics was equal to the topic we chose manually. Table 2 shows the topic accuracies of our model. Overall, our model works well, with an average accuracy well above 80%.

Journals without topics were much shorter in length. The average journal length of a journal with no topic was 114 characters, while one topic was 142 and two topics was 185. Manual inspection confirmed that these journals indeed did not contain any topic more than 70% of the time. Instead, they mostly contained sentiment that was already available from mood labels.

We conclude this section by remarking that we analyzed activity around significant events such as the 2016 American Election. We did not find statistically significant anomalies in topics mentioned since the topics we derived are mostly related to day-to-day activities.

Table 2. Accuracy of our topic model per topic.

Topic	Accuracy (%)
Work	88
Love	63
School	98
Sleep	72
Sickness	83
Missing someone	88
Family	83
Career & Finances	88
Dinner	93
Physical pain	75
Homework	98
Anxiety/depression	98
School (Activities)	85
Meals	83

5   Results

This section presents the results of our analysis. As a reminder, the input consists of: (1) the number of Reddit users subscribed to various mental-health-related subreddits and (2) journals from the journalling app, each labeled with a timestamp, mood (entered by the user), visibility (public vs. private; set by the user), and up to two topics (assigned by our topic modelling algorithm).

5.1   Frequent Mental Health Topics Across Reddit, Public Journalling and Private Journalling

We begin by comparing commonly subscribed topics on Reddit to common topics discussed on the journalling app both privately and publicly. Table 3 shows the most subscribed mental health related subreddits. Table 4 lists various statistics for the 14 topics we identified in the journalling app, including:

- Happiness percentage, corresponding to the percentage of journals whose associated mood was "happy".
- Publicness percentage, corresponding to the percentage of journals whose visibility was set to public.

Table 3. Mental health related communities on Reddit with the most subscribers.

Subreddit	Subscribers
Depression	174 k
Anxiety	110 k
ADHD	70 k
Suicide watch	53 k
Stop smoking	46 k
Mental health	26 k
Aspergers	22 k
Dating	21 k
Career Guidance	21 k
BPD	17 k
Bipolar reddit	16 k
OCD	12 k
Sleep	12 k
Eating disorders	9 k
Insomnia	9 k
Alcoholism	8 k
High school	4 k
Family	2.5 k

Table 4. Topic statistics for the journalling app.

Topic	Happiness (%)	Publicness (%)	Journals (1000s)	Users (1000s)	Average length
Dinner	83	21	20	7	143
Meals	80	26	8	4	100
School (Activities)	68	30	16	6	136
Work	64	31	86	21	147
Sickness	36	32	23	9	118
School	58	32	39	12	140
Homework	63	32	11	4	124
Physical pain	38	33	16	7	124
Family	66	34	23	10	155
Missing someone	43	35	18	8	142
Sleep	43	36	35	14	135
Career & finances	57	37	19	8	142
Love	70	38	58	17	154
Anxiety/depression	38	42	17	7	147

- Number of journals per topic (1000s).
- Number of users who posted at least one journal on the given topic (1000s).
- Average journal length per topic.

While scanning for health-related communities on Reddit, we immediately noticed that physical health (exercise, weight loss) is a much larger theme compared to the journalling dataset. This is likely since there are other apps for tracking exercise. On the other hand, communities focused on mental health were relatively small given Reddit's large user base. Self-identified depression was the largest subreddit focused on a mental health condition, which in the journalling dataset was also a common topic. Additionally, Reddit includes smaller communities, such as "High School", "Sleep" and "Family" that correspond to important topics found in the journalling dataset. Notably, people with ADHD formed a very large community on Reddit, which was not a major theme in the journalling dataset and which could be a unique dataset for researchers interested in ADHD.

On Reddit, sleep-related communities are very small while in the journalling dataset it is a major theme. Sleep is a daily need that is critical to mood, which is what the journalling app is designed to track. Sleep is the third most common topic, and, as discussed later, it has a relatively negative sentiment. Based on manual inspection of a random subset of public journals, mentions of sleep are not mainly related to insomnia. Instead, we found that most mentions of sleep include commentaries on the quality of sleep, looking forward to go to sleep due to exhaustion, and (non-chronic) lack of sleep.

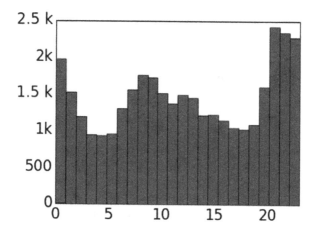

Fig. 5. Time of day when sleep was mentioned, summed across 2016.

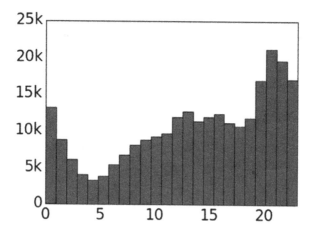

Fig. 6. Time of day when topics besides sleep and dinner were mentioned, summed across 2016.

To further understand how users are logging their sleep, we analyzed the timing of journals that mentioned sleep. Figures 5 and 6 show the times of day that sleep and non-sleep related journals, respectively, were written across all journals posted in 2016. Sleep was uniquely mentioned in the mornings, whereas all other topics followed a very similar distribution ("Dinner" was mentioned later in the day than other topics and was removed for clarity). In agreement with our manual inspection, sleep is mentioned before common hours of sleep and in the morning after waking up.

In addition, Reddit's community does not appropriately address specific issues that are affecting people in the journalling dataset, including family and school-related stress. Also, while a large subreddit exists for career advice, it does not specifically target job-related stress and workplace conflicts that are mentioned in the journalling dataset.

5.2   Analysis of Journal Moods and Visibility

Overall, one third of all journals are public. Based on Table 4, we find that social media has a gap in its ability to fulfill our social needs when expressing day to day activities. In particular, "Dinner" and "Meals" are topics that are shared (set to public) less than 30% of the time. Based on manual inspection of a random sample of public journals, those labelled with the topic "Dinner" tend to be about dates and family gatherings. On the other hand, "Meals" are generally short journals that are used to track how much was eaten and whether it was healthy or not. By creating a private medium, the journalling app helps people reflect upon these moments.

On the other hand, more public topics which were shared 35% or more of the time were "Missing Someone", "Sleep", "Career & Finances", "Love" and "Anxiety/Depression". Anxiety and depression are talked about the most publicly, which shows that users are aware of and comfortable sharing their mental state on the journalling app. In comparison, these topics are not usually found on traditional social media since there is a stigma around them.

Table 4 also contains the average mood of each topic, as labeled by users. While most topics are generally quite happy, there are some that are unexpectedly sad. Most surprisingly, "Sleep" is just as negative as "Missing Someone", with only 43% of journals happy, compared to the average happiness across the dataset of 60%. "Dinner" and "Meals" were especially happy, which also happened to be the most private topics.

6   Conclusions

In this paper, we used text mining to analyze a unique dataset of public and private journals in order to understand public mental health. We uncovered core themes affecting users. Based on user-labeled moods, we analyzed sentiment, revealing that the most private topics had the most positive mood. Despite being a very low mood topic, anxiety and depression were frequently publicly shared, showing the stigma around these issues can be mitigated in an anonymous environment.

By comparing public and private journals, we determined which topics are shared more than others, identifying new themes not available in currently analyzed social media. Routine topics such as eating meals are kept private by users. Across the dataset, most journals and topics were mostly private, suggesting that traditional social media cannot fulfill the need to express emotions during these moments.

We also compared the journalling app to Reddit, another service for anonymous sharing. We found that mental health topics such as family, school and work-related issues were missing from Reddit, perhaps because people are uncomfortable discussing these issues in a public forum, even anonymously. We believe there is an unfilled need for this user base. Future social media services may wish to offer a place to talk about these problems and make people comfortable enough to express emotions about them publicly.

An interesting finding is that sleep was a critical theme in the journalling dataset. This topic was frequently mentioned before and after sleeping. Sleep had an unexpected negative sentiment that is comparable with the topic of missing someone. Sleep is a daily activity that has a large impact on mood and is impacted by external factors such as stress. Hence, sleep monitoring data is critical for understanding public mental health.

In future work, we plan to collect more data to analyze issues related to sleep in more detail. For example, Twitter data offers a chance to study sleep patterns in users who post daily.

References

1. Bird, S., Klein, E., Loper, E.: Natural Language Processing with Python. O'Reilly Media, Sebastopol (2009)
2. Coppersmith, G., Dredze, M., Harman, C., Hollingshead, K.: From ADHD to SAD: analyzing the language of mental health on Twitter through self-reported diagnoses. In: Proceedings of the 2nd Workshop on Computational Linguistics and Clinical Psychology: From Linguistic Signal to Clinical Reality, pp. 1–10 (2015)
3. Coppersmith, G., Harman, C., Dredze, M.: Measuring post traumatic stress disorder in Twitter. In: Proceedings of the Eighth International AAAI Conference on Weblogs and Social Media (ICWSM), pp. 579–582 (2014)
4. Deziel, M., Olawo, D., Truchon, L., Golab, L.: Analyzing the mental health of engineering students using classification and regression. In: Proceedings of the 6th International Conference on Educational Data Mining (EDM), pp. 228–231 (2013)
5. Diederich, J., Al-Ajmi, A., Yellowlees, P.: Ex-ray: data mining and mental health. Appl. Soft Comput. **7**(3), 923–928 (2007)
6. Harman, G., Coppersmith, M., Dredze, C.: Quantifying mental health signals in Twitter. In: Proceedings of the Workshop on Computational Linguistics and Clinical Psychology: From Linguistic Signal to Clinical Reality (2014)
7. Hossain, H.S., Roy, N., Khan, M.: Sleep well: a sound sleep monitoring framework for community scaling. In: 2015 16th IEEE International Conference on Mobile Data Management (MDM), vol. 1, pp. 44–53 (2015)
8. Hoyer, P.O.: Non-negative matrix factorization with sparseness constraints. J. Mach. Learn. Res. **5**, 1457–1469 (2004)
9. Kuppermann, M., Lubeck, D.P., Mazonson, P.D., Patrick, D.L., Stewart, A.L., Buesching, D.P., Filer, S.K.: Sleep problems and their correlates in a working population. J. General Internal Med. **10**(1), 25–32 (1995)
10. Loper, E., Bird, S.: NLTK: the natural language toolkit. In: Proceedings of the ACL-02 Workshop on Effective tools and Methodologies for Teaching Natural Language Processing and Computational Linguistics, vol. 1, pp. 63–70 (2002)
11. Luxton, D.D., June, J.D., Kinn, J.T.: Technology based suicide prevention current applications and future directions. Telemed. eHealth **17**(1), 50–54 (2011)

12. Pedregosa, F., Varoquaux, G., Gramfort, A., Michel, V., Thirion, B., Grisel, O., Blondel, M., Prettenhofer, P., Weiss, R., Dubourg, V., Vanderplas, J., Passos, A., Cournapeau, D., Brucher, M., Perrot, M., Duchesnay, E.: Scikit-learn: machine learning in Python. J. Mach. Learn. Res. **12**, 2825–2830 (2011)
13. Perlis, R., Iosifescu, D., Castro, V., Murphy, S., Gainer, V., Minnier, J., Cai, T., Goryachev, S., Zeng, Q., Gallagher, P., et al.: Using electronic medical records to enable large-scale studies in psychiatry: treatment resistant depression as a model. Psychol. Med. **42**(1), 41–50 (2012)
14. White, M., Dorman, S.M.: Receiving social support online: implications for health education. Health Educ. Res. **16**(6), 693–707 (2001)

Clinical Data Analytics

Effects of Varying Sampling Frequency on the Analysis of Continuous ECG Data Streams

Ruhi Mahajan, Rishikesan Kamaleswaran, and Oguz Akbilgic[✉]

University of Tennessee Health Science Center, Memphis, TN 38103, USA
oakbilgl@uthsc.edu

Abstract. A myriad of data is produced in intensive care units (ICU) even for short periods of time. This data is frequently used for monitoring patient's immediate health status, not for real-time analysis because of technical challenges in real-time processing of such massive data. Data storage is also another challenge in making ICU data useful for retrospective studies. Therefore, it is important to know the minimal sampling frequency requirement to develop real-time analysis on ICU data and to develop a data storage plan. In this study, we have applied the Probabilistic Symbolic Pattern Recognition (PSPR) method in Paroxysmal Atrial Fibrillation (PAF) screening problem by analyzing electrocardiogram signals at different sampling frequencies varying from 128 Hz to 8 Hz. Our results show that using PSPR method, we can obtain a classification accuracy of 82.67% in identifying PAF subjects even when the test data is sampled at 8 Hz frequency (73.33% for 128 Hz). This classification accuracy drastically improved to 92% when other descriptive features were used along with PSPR features. The PSPR's PAF screening ability at low sampling frequency indicates its potential for real-time analysis and wearable embedded computing applications.

Keywords: Sampling frequency · Electrocardiogram (ECG) · ICU data · Paroxysmal atrial fibrillation · Probabilistic Symbolic Pattern Recognition

1 Introduction

The intensive care units (ICUs) have frequently been identified as the information intensive care unit due to the great variety of devices available to produce data about the patient's immediate health status. One of the major components of this data collection process involves physiologic medical devices. Medical devices at the bed-side usually sample data at a high sampling rate. For instance, an electrocardiogram (ECG) sensor, samples data at 1 kHz over two leads to produce waveform signals of the patient's real-time cardiovascular function. The ventilator samples in that same range to produce rich information about the patient's respiration. Other medical devices sample data in similar methods to identify the patient's current physiologic status. Each of these devices produces significant volumes of data that are usually truncated to support existing algorithmic design, standardize signals with varying sample frequencies, and

© Springer International Publishing AG 2017
E. Begoli et al. (Eds.): DMAH 2017, LNCS 10494, pp. 73–87, 2017.
DOI: 10.1007/978-3-319-67186-4_7

conform to the space limitations of the electronic medical record (EMR), or the database infrastructure that supports them.

One of the common methods to reduce the data requirement is to use signal decimation [1]. Decimation has been utilized by large data warehouses, such as the MIMIC (Multi-parameter Intelligent Monitoring in Intensive Carte) – III database to standardize heterogeneous signals across the database, in addition to optimizing the storage of physiologic data [2]. CapnoBase meanwhile presents data that was up-sampled to maintain homogeneity across the capnogram and the photoplethysmo-gram [3]. While these systems present opportunities for making large volumes of physiologic data available for analysis, there is a need to identify the influence various sampling frequencies might have on algorithms that depend on them.

In this paper, we evaluated the performance of a novel Probabilistic Symbolic Pattern Recognition (PSPR) approach at various sampling frequencies of ECG, to identify individuals at risk of Paroxysmal Atrial Fibrillation (PAF). Atrial fibrillation is one of the major and ever growing cardiovascular problems which affects about 2–3% of the population of the United States. To accurately identify PAF during normal sinus rhythm is a challenging problem because PAF episodes are asymptomatic and often occur for a brief period. Timely detection of PAF is important as longer episodes increase the risk of ischemic stroke [4] and can evolve into permanent atrial fibrillation within four years [5]. While there are various statistical models available in the liter-ature to screen PAF [6–9], some with an accuracy up to 82% [10], none of the existing models are suitable for the embedded computing on wearable devices. This is mainly due to memory constraints that limit the processing of data with high sampling rates e.g., 256, 500 Hz.

In our previous study, we have applied the PSPR algorithm to classify the PAF data [11]. The main objective of this study is to identify a reasonably low sampling rate of ECG recordings that can still differentiate normal subjects from PAF subjects. It is shown in this study that PSPR algorithm can accurately screen PAF even when the normal sinus rhythm is sampled at 8 Hz.

This manuscript is organized as follows. Section 2 introduces the concept of PSPR algorithm, followed by its implementation protocol on the PAF database, other feature extraction techniques, and classification procedure in Sect. 3. The key results are mentioned in Sect. 4, some findings are discussed in Sect. 5, and concluding remarks in Sect. 6.

2   Probabilistic Symbolic Pattern Recognition (PSPR)

The PSPR technique learns pattern transition behaviors in sequential series represented by n_s unique symbols [11]. When applied to a stationary time series, the time series is first discretized using user defined thresholds generally based on the distribution of the series (i.e. normal, skewed, etc.). The commonly used Piecewise Average Approximation (PAA) [12, 13], Symbolic aggregate approximation (SAX) [14], Extended SAX [15], etc. based techniques can be also applied for discretizing series. Figure 1 shows an example of a symbolically discretized one second ECG excerpt using the discretization rules in Table 1.

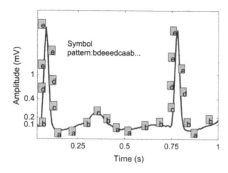

Fig. 1. Five-symbols quantile based discretization of a sample ECG recording (discretization rule as specified in the Table 1).

Table 1. Five-symbol quantile based discretization rules for a sample ECG recording.

Symbol alphabet	Range
a	[-∞ 0.10]
b	[0.10 0.20]
c	[0.20 0.40]
d	[0.40 1.0]
e	[1.0 ∞]

After discretizing series, the PSPR algorithm finds the joint occurrence of observed patterns of up to length n_p that is followed by a single symbol. For example, a two-symbols pattern transition *ab* followed by *c* or a three-symbol pattern transition *bdc* followed by *a*. By observing the frequency of occurrence of such pattern transitions, pattern transition probabilities (PTP) can be calculated. Consider an example of discretized series, $S = \{ababbbaccbbabcabacc\}$ defined using three-symbol {a, b, c} i.e., $n_s = 3$. For $n_p = 2$, PTP_2 can be calculated based on two-symbol pattern transitions (refer Table 2). According to PTP values, if we observe either *ca* or *cb*, it is likely that the next symbol will be *b*. However, if we observe *ba*, symbols *b* and *c* are equiprobable. Therefore, using PSPR, any sequential series can be converted into a

Table 2. PTP_2, a two-symbol pattern transition probability for an example series, S.

Symbol pattern	PTP_2		
	a	b	c
ab	0.5	0.25	0.25
ac	0	0	1
ba	0	0.5	0.5
bb	0.5	0.5	0
bc	1	0	0
ca	0	1	0
cb	0	1	0

matrix of PTPs. The optimal selection of n_s and n_p are important to balance the computational complexity vs. information needed.

Further, according to the PSPR framework, two sequential series can be compared by computing the similarity between their PTPs. If series are similar, then the pattern transition similarity (PTS) matrix would be close to 0 or exactly 0 when series are the same. The similarity between PTPs of series can be computed using different metrics like Euclidean distance, Mahalanobis distance, cosine similarity, etc. [16]. In this study, however, we computed similarity by calculating Euclidean distance between the PTPs of series PTP_{ik} and PTP_{jk} as:

$$PTS_{i,j} = \sum_{k=1}^{N_p} eucdist(PTP_k^i, PTP_k^j) \tag{1}$$

where $k = 1,2...N_p$, with $N_p = max(n_p)$. To simplify the calculations, PTP matrix is zero-padded to fill any missing pattern transition probabilities.

Clustering of m discretized series in PSPR can be thereby performed by computing and comparing PTS for all m series, which will result in an $m \times m$ symmetric matrix with $PTS_{i,j}$ being the distance between the ith and j^{th} series. Thus, by utilizing the probabilistic transition behaviors in an arbitrary length series, underlying patterns can be characterized and clustered using the novel PSPR approach. In comparison to the conventional symbolic representation techniques, some of the unique features of PSPR are:

(1) It can compare symbolic series of different length.
(2) It does not require time stamp to compare series.
(3) It does not compare series based on their co-movement, instead, it compares series based on their reaction to an observed set of series.
(4) It retrieves unique features of the series without any information loss.

Figure 2 represents the block diagram of the PSPR algorithm used for supervised classification of PAF and normal groups in this study. The I/P series in this study are the ECG recordings (sampled at 128, 64, 32, 16, and 8 Hz) from both normal subjects and subjects with PAF condition.

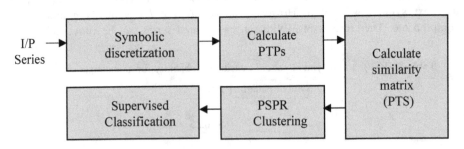

Fig. 2. Block diagram of the PSPR algorithm to identify normal healthy subjects and subjects with PAF condition.

3   Procedure

This section presents details on the dataset, decimation procedure, implementation of the PSPR algorithm, other feature extraction techniques, and supervised classifiers used to distinguish normal subjects and subjects prone to PAF.

3.1   PAF Prediction Challenge Database

The PSPR algorithm was applied to the training set of the PAF prediction database from the PhysioNet and Computers in Cardiology 2001 challenge [17, 18]. This database contained excerpts of half-hour digitized ECG recordings (sampled at 128 Hz with 12-bit resolution) from 75 subjects. Among all subjects, 50 subjects didn't have any PAF history and so they were considered to be in the normal (control) group, annotated as NN in this study. For the remaining 25 subjects who had PAF (considered in the PAF group and annotated as PAF_N), their ECG recordings were at least 45 min distant from any PAF episode. The database also included five minute ECG recordings during an episode of PAF for the same 25 subjects. These recordings have been annotated as PAF_E and were later used in this study for comparison with NN and PAF_N recordings.

3.2   Signal Decimation

To investigate the impact of low sampling rate on the PSPR algorithm, ECG signals of both NN and PAF_N groups were decimated by a factor of two, four, eight and sixteen. The decimated ECG signals were therefore obtained at a sampling frequency, $f_s = 64$, 32, 16 and 8 Hz, respectively. To avoid aliasing, a zero phase 7th order digital low pass IIR filter was applied on ECG signals before down sampling using MATLAB (MathWorks, CA, USA). The cut-off frequency of this anti-aliasing filter was determined to satisfy the Nyquist sampling theorem as:

$$f_c = f_s / (2 * factor) \qquad (2)$$

where *factor* is the ratio of decimation for each decimated ECG signal.

3.3   Discretization and PSPR Clustering

To apply the PSPR algorithm, each continuous ECG recording was discretized using five-symbol alphabets {a, b, c, d, e}. As the ECG database was skewed in nature, quantile-based thresholds were applied on the recordings as mentioned in Table 1.

In the next step, PTPs for each discretized recording were computed. The maximum pattern length, n_p for the analysis was chosen to be seven (details are mentioned in the result section). The pattern transition similarity matrix (using a *Euclidean* distance metric) of size 75 × 25 was then formed, which compared PTPs of NN, PAF_N and PAF_E recordings with each other. Lastly, PTP distances in the similarity matrix were averaged to generate a 75 × 1 vector depicting average distance of NN and PAF_N PTPs from the PAF_E PTPs. This vector was further evaluated to find any significant differences in the median between NN and PAF_N group PTPs, it has been discussed in detail in the result section.

3.4   Predictive Models

We used a simple binomial logistic regression analysis to determine if the average distance from PAF_E could automatically classify individuals into NN and PAF_N groups. The main objective of a logistic regression model is to predict a logit transformation of the probability of the presence of a predictor, i.e. for a given output variable Y, simple logistic regression models the conditional probability $P(Y = 1| X = x)$ as a linear function of x. Mathematically, a logistic model can be expressed as:

$$\log \frac{p(x)}{1 - p(x)} = \beta_o + x * \beta \qquad (3)$$

where p is the probability of the presence of the predictor x and β_o, β are the coefficients of the regression equation. By choosing a threshold, p_{cutoff}, for predicted probabilities, p, logistic regression model, Y, therefore gives us the linear classifier, e.g., Y = 1 when $p > p_{cutoff}$ and Y = 0 otherwise. We applied a logistic regression classifier on average distance from PAF_E to classify individuals into categories; normal (Y = 0) or PAF (Y = 1).

To avoid overfitting and to validate the classifier, we used a five-fold cross validation technique in which the PTS matrix was partitioned into five equal sized sub-matrices. We used 80% data to train the logistic model and then tested it on remaining 20% data. The results from each fold were averaged to produce a single estimation. This classification process was applied to each PTS matrix obtained with ECG recordings at f_s = 8, 16, 32, 64, and 128 Hz.

Later, in the study, we obtained better results with a Support vector machine (SVM) classifier that performed classification by constructing an optimal hyperplane to separate data of the NN and PAF_N group. The features obtained were not linearly separable so we used a quadratic kernel (larger degree kernel tend to overfit) to generate the hyperplane. Mathematically, the kernel function can be expressed as:

$$K(x, y) = (x^T y + 1)^2 \qquad (4)$$

where x and y are the feature vectors in the input space. The generalizability of the SVM classifier was evaluated with a five-fold cross validation technique.

3.5   Feature Selection

To generalize the classification model and scrutinize significant features to be used for the classification, we also evaluated the performance of the PSPR algorithm by using a sequential forward feature selection (SFFS) approach. The SFFS method selects a subset of features that best predict the data by sequentially selecting features until there is no improvement in the prediction. In this study, SFFS selects between pattern transitions probabilities for different symbols i.e. between PTP_1- PTP_7. The feature selection module in the PSPR algorithm also helps to identify the optimum length of pattern transitions (n_p) in a series that can distinguish normal subjects from other subjects prone to PAF.

4 Results

This section explains the key findings obtained with the application of the PSPR algorithm on the PAF database. The results are also compared with the existing techniques in the literature.

4.1 Evaluation of Effect of Symbol Size (n_s)

The symbol size, n_s is a user-defined parameter that can affect the performance of PSPR. A large n_s results in more computational time and complexity, whereas, a small n_s may not be sufficient to retrieve the underlying information of the input series. We, therefore, analyzed different values of n_s = 4, 5, and 6 to find the optimum value which results in a high classification rate for the PAF database. Each value of n_s is analyzed for different sampling frequencies (f_s = 8, 16, 32, 64, and 128). Figure 3 depicts that with n_s = 5, maximum accuracy is achieved to differentiate between NN and PAF_N groups.

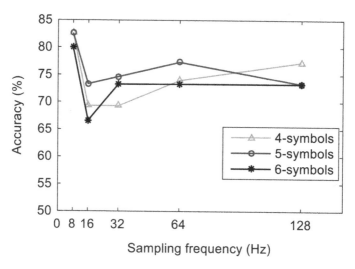

Fig. 3. Classification accuracy at different sampling frequencies for various symbol size.

4.2 Determination of Symbol Pattern Length (n_p)

As explained before using the PSPR method, PTP matrices will be calculated for each series up to n_p-symbol pattern transitions. Here, n_p is a very critical parameter of our method. Optimal determination of n_p is important to balance computational cost vs. information modeled. Like n_s, when n_p is small, the method will not truly learn the pattern transition behavior of the series. On the other hand, setting n_p too high will be computationally expensive and result in a sparse transition matrix. To select the optimal value for n_p, we used a criterion called the Sparsity Index (SI), mathematically computed as:

$$SI = \frac{No.\,of\;observed\;n_p}{No.\,of\;possible\;n_p} \tag{5}$$

For illustration purpose, consider a discretized series, $S = \{abcaabc\}$, defined using three symbols. In this series, alternative one-symbol patterns (a, b, or c) and SI = 3/3 = 100% is observed. However, there are only six (ab, bc, ca, aa, ab, bc) of the nine possible two-symbol patterns observed in S i.e. SI = 6/9 = 33%. Similarly, for three-symbol patterns, the SI would be calculated as 5/27 = 19%. Obviously, the SI value is a decreasing function of the pattern length. Therefore, we limit adding new PTP matrices when the calculated SI is below a given threshold value. In this study, we used a threshold of 0.000001, which resulted in $n_p = 7$.

4.3   PSPR Clustering

The PSPR algorithm to screen subjects with PAF was evaluated at various sampling frequencies. The algorithm symbolically discretized all ECG recordings from the NN, PAF_N, and PAF_E groups using five-symbol. Then, up to seven PTP matrices (PTP_1, $PTP_2,...,PTP_7$) were computed which thereby formed a similarity matrix comparing the

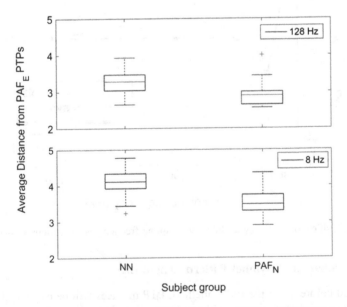

Fig. 4. Box plot depicting average of seven PTP distances of NN and PAF_N groups from PAF_E group when ECG recordings were sampled at 128 and 8 Hz. Lower distances on the y-axis represent more similarity with PAF_E ECG recordings.

NN and PAF_N recordings to PAF_E. The lower average PTP distance in the box plot in Fig. 4 suggests that PAF_N ECG recordings were more similar to the PAF_E ECG recordings, irrespective of the sampling frequency. Although we have not included box plot for other sampling frequencies, the trends were found to be similar.

4.4 PAF Classification Without Feature Selection

As mentioned earlier, logistic regression based classifier was implemented on PTS matrix obtained using ECG recordings at various sampling frequencies. The average classification accuracies from five-fold cross-validation are tabulated in Table 3 for the training and test data at different f_s. It can be observed that the test data accuracy increases with the decrease in sampling frequency for five-symbols PSPR. This indicates that PSPR can efficiently detect subtle long-term differences prevailing in ECG patterns. This also suggests that information at 8 Hz is sufficient to train PSPR model to screen subjects with PAF condition. The model was trained on a small dataset, yet it could correctly discriminate NN and PAF_N groups with an accuracy of $\sim 83\%$, slightly higher than what was reported in the literature. Table 4 represents the approximate computational time taken by the three important stages of the PSPR method. It can be noticed that most of the time is consumed while computing the PTS matrix which compares the PTPs of each series. Time was calculated by the stop watch timers in MATLAB running on a machine with 2×2.4 GHz 6-core Intel processor.

Table 3. Five-fold average classification accuracy of a logistic regression classifier obtained with PSPR features at different sampling frequencies of ECG recordings.

Sampling frequency, f_s (Hz)	Accuracy (%)	
	Training data	Test data
128	76.67	73.33
64	81.67	77.33
32	85	74.67
16	85.33	73.33
8	84.33	82.67

Table 4. Elapsed time for the execution of PSPR algorithm on ECG recordings sampled at 8 Hz from NN and PAF_N groups.

PSPR computation stage	Computation time
Five-symbol symbolic discretization	0.30 s
Up to seven PTP calculation (PTP_1.... PTP_7)	4.6 min.
PTS calculation using Euclidean distance metric	10 min.

Furthermore, we calculated C-statistics i.e., area under the receiver operating characteristic (AUC) curve for our model at all test f_s of ECG recordings. The AUC as observed from Fig. 5 was 0.816 and 0.768 for the test data at 8 Hz and 128 Hz, respectively. Thus, we conclude that with PSPR we can classify individuals in NN and PAF_N groups by scanning their normal sinus rhythm sampled at 8 Hz.

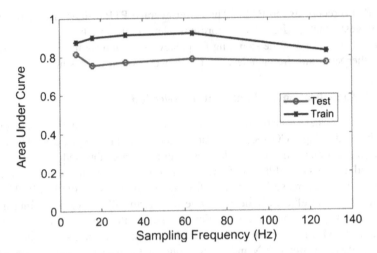

Fig. 5. Area under the ROC curve at various sampling frequencies obtained with the logistic regression classifier. The original database was split into test and training datasets with five-fold cross validation technique.

4.5 Statistical Test

To test the equality of the median of the NN and PAF_N average distances, a non-parametric Mann-Whitney U test was used. Unlike the t-test it does not need the assumption of normal distributions of data. For both NN and PAF_N series, the null hypothesis of equal proportions was rejected with a very low p-value (< 0.001) for all test-sampling frequencies. This implies that discretization and the values of the features extracted by PSPR method are significantly different series for NN and PAF_N groups. In other words, we can say that PSPR based features obtained from ECG recordings are discernible enough to separate subjects in two groups/classes.

4.6 Improved PAF Classification with Feature Selection and Additional Features

In previous section, we outlined the classification accuracy in detecting PAF using only features extracted by PSPR approach. To improve the classification performance of the predictive model, we evaluated and computed a more descriptive features like mean, median, kurtosis, skewness, and range of 8 Hz sampled ECG recordings. Also, we trained the classification model using only the significant features selected by the SFFS method. The discretization and clustering process remains the same as described in the previous sections. We found that among all PSPR features, PTP_7 and PTP_6 provide most significant information to differentiate PAF_N group from the normal group. To get a better understanding, Fig. 6 depicts the average similarity distance between NN and PAF_N group using PTP_7. Our preliminary analysis suggested that using eight features (six descriptive and two PSPR) and with an SVM classifier, it is possible to screen subjects with PAF condition with an accuracy of 92%. A red marker in the ROC curve

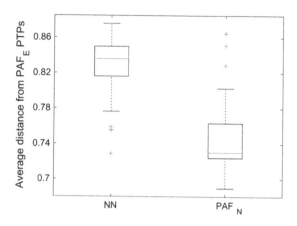

Fig. 6. Box plot of the average distance of NN and PAF_N group from PAF_E PTPs using only PTP_7 (ECG signals at $f_s = 8$ Hz).

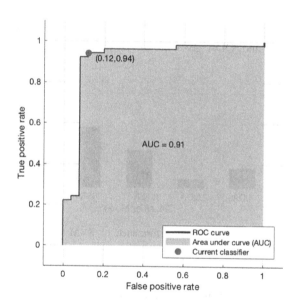

Fig. 7. ROC curve of final predictive model with the SVM classifier implemented using PSPR and descriptive features from ECG recordings sampled at 8 Hz. (Color figure online)

in Fig. 7 suggests that the classifier has a true positive rate of 0.94 and false positive rate of 0.12, resulting in an overall AUC = 0.91. The average five-fold cross-validation classification statistics with this final model are reported in the confusion matrix in Table 5.

Table 5. Confusion matrix for the final SVM based classification model using eight features from the PAF database sampled at 8 Hz.

	Predicted		
Actual	Normal	PAF	
Normal	47	3	Specificity = 94%
PAF	3	22	Sensitivity = 88%
	Negative Predictive Value = 94%	Positive Predictive Value = 88%	

4.7   Comparison with Other Classifier Models

To find the best predictive model we analyzed various classification models like decision trees, linear discriminant analysis (LDA), quadratic discriminant analysis (QDA), logistic regression (LR), k nearest neighbor (kNN, where k = one), and ensemble of bagged trees along with the SVM classifier. The five-fold cross validation accuracies obtained from each model are plotted in Fig. 8. It can be noted that an SVM classifier resulted in the highest accuracy of 92%. SVM finds a hyperplane using the quadratic kernel that maximizes the margin between data points of NN and PAF_N groups. On the other hand, QDA gives the lowest classification accuracy of 80%.

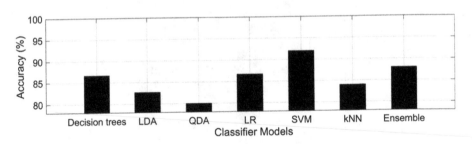

Fig. 8. Comparison of the classification performance of SVM with other commonly used classifiers using the same dataset.

4.8   Comparison with Existing Techniques on PAF Database

To the best of our knowledge, this is the first study to report the PAF screening problem using ECG data sampled at a low frequency. Most of the studies in the literature have used the PAF database recordings at 128 Hz sampling rate. Table 6 mentions the accuracy obtained by other groups on the same 2001 PAF challenge database used in this study. As observed, the highest accuracy to classify subjects prone to PAF is 82% (on the test data). However, in this study, with PSPR features along with some descriptive features from the ECG recordings the classification accuracy can be increased up to 92%.

Table 6. Classification accuracies reported on the 2001 PAF database challenge.

Related studies	Test data accuracy
G Schreier *et al.* [10]	82%
W Zong *et al.* [8]	80%
R Sweeney *et al.* [20]	74%
C Maier *et al.* [9]	72%
C Marchesi *et al.* [21]	70%
Proposed study	**92%**

5 Discussion

The aim of this study was to evaluate how well PSPR based approach can screen PAF if the normal sinus rhythm is sampled at a very low rate. The motivation to find a reasonable low sampling rate is the increased interest towards miniature, wearable networked health monitoring devices which have bandwidth constraints.

We have reported classification results on the training PAF database of PhysioNet (NN = 55, PAF_N = 25 and PAF_E = 25). We would have preferred to test our classifier on the test PhysioNet PAF database but the *autoscorer* webpage is no longer available. Nevertheless, our analysis on the small dataset indicates that it is possible to capture underlying fingerprints of PAF in a normal ECG. It is also worth to note that the PAF_N ECG recordings were at least 45 min distant from any PAF episode; even then PSPR could identify the underlying similarity. Also, unlike commonly used SAX/ ESAX methods, PSPR could compare ECG series of different lengths (5 min PAF episode vs. 30 min NN and PAF_N).

We have also evaluated PSPR framework using symbol discretization rule on the R-R intervals at test sampling frequencies. The R-R intervals for NN and PAF_N ECG recordings were obtained using the conventional Pan-Tompkins algorithm [19]. However, the classification results were not comparable, hence were not included in this paper. It means that the fingerprints of PAF can be found without taking R-R intervals into account. In another study, we applied PSPR technique to identify subjects with congestive heart failure condition [22]. Using PSPR along with some descriptive measures on the R-R interval series we obtained a classification accuracy of 99.5%.

The results of this study suggest that the PSPR model can efficiently identify underlying features of the ECG recording even when the recording has a low temporal resolution. It is worthwhile to mention that we obtained high classification accuracies using only one predictor- one channel ECG data, simple descriptive features, and a very small training dataset. We expect that by including more predictors like vitals, demographics, medication use, etc. would further increase the accuracy of results.

6 Conclusion

In this study, the Probabilistic Symbolic Pattern Recognition (PSPR) framework was applied in PAF screening for the given ECG at various sampling frequencies, 8,16, 32, 64, and 128 Hz. Using the features extracted by PSPR and some descriptive features, a

SVM classifier was implemented for supervised classification of normal and PAF subjects. The five-fold cross validation results suggest that the PSPR based predictive model along with descriptive metrics can diagnose PAF with an accuracy, sensitivity, and specificity of 92%, 88%, and 94%, respectively even when the given ECG recordings are sampled as low as 8 Hz. These results pave a path to implement PSPR based model in the ambulatory setting involving networked devices that can be used to screen, diagnose, and generate early warnings in health monitoring systems without significant bandwidth requirements.

References

1. Milic, L.: Multirate Filtering for Digital Signal Processing: MATLAB Applications: MATLAB Applications. IGI Global (2009)
2. Johnson, A.E.W., et al.: MIMIC-III, a freely accessible critical care database. Sci. Data **3**, 1–9 (2016)
3. Karlen, W., et al.: Multiparameter respiratory rate estimation from the photoplethysmogram. IEEE Trans. Biomed. Eng. **60**(7), 1946–1953 (2013)
4. Abdul-Rahim, A.H., Lees, K.R.: Paroxysmal atrial fibrillation after ischemic stroke: how should we hunt for it? Expert Rev. Cardiovasc. Ther. **11**(4), 485–494 (2013)
5. Al-Khatib, S.M., et al.: Observations on the transition from intermittent to permanent atrial fibrillation. Am. Heart J. **140**(1), 142–145 (2000)
6. Lynn, K.S., Chiang, H.D.: A two-stage solution algorithm for paroxysmal atrial fibrillation prediction. Comput. Cardiol. **28**, 405–407 (2001)
7. Yang, A.C.C., Yin, H.W.: Prediction of paroxysmal atrial fibrillation by footprint analysis. Comput. Cardiol. **28**, 401–404 (2001)
8. Zong, W., Mukkamala, R., Mark, R.G.: A methodology for predicting paroxysmal atrial fibrillation based on ECG arrhythmia feature analysis. Comput Cardiol. **28**, 125–128 (2001)
9. Maier, C., Bauch, M., Dickhaus, H.: Screening and prediction of paroxysmal atrial fibrillation by analysis of heart rate variability parameters. Comput. Cardiol. **28**, 129–132 (2001)
10. Schreier, G., Kastner, P., Marko, W.: An automatic ECG processing algorithm to identify patients prone to paroxysmal atrial fibrillation. Comput. Cardiol. **28**, 133–135 (2001)
11. Akbilgic, O., Howe, J.A., Davis, R.L.: Categorizing atrial fibrillation via symbolic pattern recognition. J. Med. Stat. Inform. **4**(8), 1–9 (2016)
12. Keogh, E., et al.: Dimensionality reduction for fast similarity search in large time series databases. Knowl. Inf. Syst. **3**(3), 263–286 (2001)
13. Yi, B.-K., Faloutsos, C.: Fast time sequence indexing for arbitrary Lp norms (2000)
14. Lin, J., Keogh, E., Lonardi, S., Chiu, B.: A symbolic representation of time series, with implications for streaming algorithms. In: Proceedings of the 8th ACM SIGMOD Workshop on Research Issues in Data Mining and Knowledge Discovery, pp. 2–11 (2003)
15. Lkhagva, B., Suzuki, Y., Kawagoe, K.: Extended SAX: extension of symbolic aggregate approximation for financial time series data representation, DEWS2006 4A-i8, vol. 7 (2006)
16. Deza, M.M., Deza, E.: Encyclopedia of Distances, pp. 1–583. Springer, Heidelberg (2009)
17. Goldberger, S.H., et al.: PhysioBank, PhysioToolkit, and PhysioNet components of a new research resource for complex physiologic signals. Circulation **101**(23), 215–220 (2000)
18. Physionet. The PAF Prediction challenge database (2001). https://www.physionet.org/physiobank/database/afpdb/Accessed 17 Oct 2016]

19. Jiapu, P., Tompkins, W.J.: A real-time QRS detection algorithm. IEEE Trans. Biomed. Eng. **BME-32**(3), 230–236 (1985)
20. Physionet. Computers in Cardiology Challenge 2001 Top scores (Final) (2001). https://www.physionet.org/challenge/2001/top-scores.shtml. Accessed 1 Jan 2017
21. Goldberger, A., McClennen, S., Swiryn, S.: Predicting the onset of Paroxysmal Atrial Fibrillation: The computers in cardiology challenge 2001. In: Computers in Cardiology 2001, pp. 113–116. IEEE (2001)
22. Mahajan, R., Viangteeravat, T., Akbilgic, O.: Improved detection of congestive heart failure via probabilistic symbolic pattern recognition and heart rate variability. Int. J. Med. Informatics (2017, in press)

Detection and Visualization of Variants in Typical Medical Treatment Sequences

Yuichi Honda[1]([✉]), Muneo Kushima[2], Tomoyoshi Yamazaki[2],
Kenji Araki[2], and Haruo Yokota[1]

[1] Department of Computer Science, Tokyo Institute of Technology,
2-12-1 Oookayama, Meguro-ku, Tokyo 152-8550, Japan
honda@de.cs.titech.ac.jp, yokota@cs.titech.ac.jp
[2] Faculty of Medicine, University of Miyazaki Hospital, 5200 Machikihara,
Kiyotake-Cho, Miyazaki-shi, Miyazaki 889-1692, Japan
jyoho_support@med.miyazaki-u.ac.jp

Abstract. Electronic Medical Records (EMRs) are widely used in many large hospitals. EMRs can reduce the cost of managing medical histories, and can also improve medical processes by the secondary use of these records. Medical workers including doctors, nurses, and technicians generally use clinical pathways as their guidelines for typical sequences of medical treatments. The medical workers traditionally generate the clinical pathways themselves based on their experiences. It is helpful for the medical workers to verify the correctness of existing clinical pathways or modify them by comparing the frequent sequential patterns in medical orders computationally extracted from EMR logs. Thinking that the EMR is a database and a typical clinical pathway is a frequent sequential pattern in the database in our previous work, we proposed a method to extract typical clinical pathways as frequent sequential patterns with treatment time information from EMR logs. These patterns tend to contain variants that are influential in verification and modification. In this paper, we propose an approach for detecting the variants in frequent sequential patterns of medical orders while considering time information. Since it is important to provide visual views of these variants so the results can be used effectively by the medical workers, we also develop an interactive graphical interface system for visualizing the results of variants in clinical pathways. The results of applying the approach to actual EMR logs in an university hospital are reported.

Keywords: Sequential pattern mining · Electronic Medical Records · Clinical pathways variant · Visualization of clinical pathways

1 Introduction

Most large hospitals use Electronic Medical Records (EMRs) to reduce the cost of managing the medical histories of patients. They speed up access to information about patients and reduce the space requirements for keeping these records.

© Springer International Publishing AG 2017
E. Begoli et al. (Eds.): DMAH 2017, LNCS 10494, pp. 88–101, 2017.
DOI: 10.1007/978-3-319-67186-4_8

Moreover, the possibility of secondary use of these records is expected to improve many aspects of medical processes. For example, frequent sequential patterns of medical orders can be extracted computationally from the EMR logs.

Medical workers including doctors, nurses, and technicians currently use *clinical pathways*. A clinical pathway is a guideline for a typical sequence of medical orders for a disease, which is traditionally generated by the medical workers themselves, based on their medical experiences. Human verification and modification of clinical pathways are time-consuming for workers. It would be helpful if medical workers could verify the correctness of the existing clinical pathways or modify them by comparing them with the frequent sequential patterns of medical orders extracted from EMR logs.

In our previous work [1], we proposed a method for extracting frequent sequential patterns from EMR logs with treatment time intervals between the medical orders. The results of experiments using actual EMR logs indicated that the frequent sequential patterns tend to contain a number of variants. The variant indicates brunched medical treatments in the clinical pathway. For example, when there are two sequences, (Surgery → Injection → Leaving Hospital) and (Surgery → Prescription → Leaving Hospital), "Injection" and "Prescription" are called the variant. These variants can affect machine-based verification and modification for supporting medical workers.

In this paper, we propose an approach of detecting the variants in the frequent sequential patterns of medical orders while considering the time intervals between them. To find meaningful variants, our method combines similar pathways and makes unshared parts of pathways variants. In addition, it is important to provide visual views of these variants so the results can be used effectively by medical workers. The readability of character-based variant results is very low. Therefore, we have developed an interactive graphical interface system for visualizing the results of variants in clinical pathways.

We applied the proposed method to actual EMR logs from a university hospital. The results indicate that our approach of variant detection and visualization is effective in supporting medical workers when verifying and modifying clinical pathways.

The remainder of this paper is organized as follows. Related research is reviewed in Sect. 2. The proposed method is described in Sect. 3. An experiment applying the proposed method is described in Sect. 4. The paper concludes in Sect. 5.

2 Related Work

There have been many studies of the secondary uses of EMRs related to clinical pathways. Wakamiya and Yamauchi proposed standard functions for electronic clinical pathways [2]. Hirano and Tsumoto proposed a method for extracting typical medical treatments by analyzing the logs stored in a hospital information system [3].

Uragaki et al. proposed a method using sequential pattern mining to extract clinical pathways. The a-priori-based frequent pattern mining algorithm [4] is well known. However, this algorithm is time-consuming with large data sets and cannot handle time intervals between items. Uragaki et al. therefore proposed T-Prefixspan [1]. This algorithm is a time-interval sequential pattern mining (TI-SPM) algorithm that can handle time intervals between items [5]. Because it is based on PrefixSpan [6], it is faster than a-priori-based algorithms. Uragaki et al. showed that more effective mining can be performed by focusing on the efficacy of the medicine. However, this approach does not process variants, so an additional capability is needed to detect clinical pathways containing variants.

Episode mining [7] is a well-known method for extracting patterns containing variants. However, this approach normally extracts such patterns from a single long sequence of data. Because we combine multiple sequences of data, this approach cannot be used for this study.

3   Method

3.1   Extracting Typical Clinical Pathways

First, we extract typical clinical pathways from actual EMRs by sequential pattern mining. As a side note, these clinical pathways do not have variants. In medical care, it is important to consider time intervals between medical treatments. In this study, we employ T-PrefixSpan [1], because this method considers time intervals between items. As in our previous study, we define medical treatments to have the four values *Class*, *Description*, *Code*, and *Name*.

- *Class* denotes the classification of a medical treatment
- *Description* denotes its detailed diagnostic record
- *Code* is the medical code that represents the unique efficacy of the medicine considered
- *Name* is the name of the medicine

For treatments without medicine, *Code* and *Name* are set to "null", to represent a blank value.

For example, assume a medical treatment designated as "injection" is described as an "intravenous injection" with a medical code "331" and the name "Lactec injection" appears in a medical log. The item is then represented in the form: (injection; intravenous injection; 331; Lactec injection). In this example, "*Code* 331" indicates "blood substitute." In another example, when the medical treatment is the "nursing task" of "changing the sheets," the item is represented in the form: (nursing task; changing the sheets; null; null).

We specify that medical treatments have four values but we execute mining by focusing on *Name*, that is, we do not use efficacy of medicines (*Code*). Efficacy is used to detect variants. This is explained in the next section.

The followings are the important definitions introduced for T-PrefixSpan [1], which we employ in this work.

Definition 1. *T-item (i, t)*

Let I be a set of items and let t be the time when an item i occurred. We define a **T-item** (i, t) as a pair of i and t.

Definition 2. *T-sequence s and O-sequence O_s*

T-sequence s is a sequence of *T-items*, which is denoted by $s =< (i_1, t_1), (i_2, t_2), \cdots, (i_n, t_n) >$. T-times that occur at the same time should be arranged in alphabetical order. Furthermore, let n be the length of T-sequence s and let an **O-sequence** of s be the sequence $O_s =< i_1, i_2, \cdots, i_n >$.

Definition 3. *Time interval TI_k*

Given a T-sequence $s =< (i_1, t_1), (i_2, t_2), \cdots, (i_n, t_n) >$, The **time interval** TI_k is defined as follows:

$$TI_k \equiv t_{k+1} - t_k \ (k = 1, 2, \cdots, n - 2, n - 1).$$

Definition 4. *T-sequential database D and O-sequential database O_D*

Given a set of T-sequences, the **T-sequential database** D is defined as follows, where the identifier *sid* of an element of D has a unique value for each sequence.

$$D \equiv \{(sid, s) | sid, s \in S\}.$$

Let an **O-sequential database** O_D be a sequential database that consists of O-sequences configured from all the T-sequences in D. Let $Size(D)$ be $Size(O_D)$, which is the number of sequences in O_D.

Definition 5. *T-frequent sequential pattern P*

Let MinSup $(0 \leq MinSup \leq 1)$ be a minimum support and let D be a T-sequential database. Given $P =< i_1, X_1, i_2, X_2, \cdots, i_{n-1}, X_{n-1}, i_n > (\forall j \ i_j$ is an item, $\forall k \ X_k$ is a set of five values: $(min_k, mod_k, ave_k, med_k, max_k))$, and we can configure a sequence $O_P =< i_1, i_2, \cdots i_{n-1}, i_n >$. We define P as a **T-frequent sequential pattern** if O_P is a frequent sequential pattern in an O-sequential database configured from D (i.e., $Sup(P) =| \{Seq \mid O_P \subseteq Seq, (sid, Seq) \subset O_D$, where sid is an identifier of $Seq\} |\geq Size(O_D) \times MinSup)$. Let O_P be the O-pattern of P. The set of five values is defined as follows:
Given all the T-sequences with O-sequences containing O_P in D, let S be one of them, where $S =< i'_1, t_1, i'_2, t_2, \cdots, i'_{m-1}, t_{m-1}, i'_m >$.
By using $j_1, j_2, \cdots, j_{n-1}, j_n$, which satisfies:

(1) $1 \leq j_1 < j_2 < \cdots < j_{n-1} < j_n \leq m$ and
(2) $i_k = i'_{j_k}, i_{k+1} = i'_{j_{k+1}}$,

we can configure sets of time intervals: $Set_{TI_1}, Set_{TI_2}, \cdots, Set_{TI_{n-1}}$, where $TI_k = t'_{j_{k+1}} - t'_{j_k}$. Moreover, in $X_k = (min_k, mod_k, ave_k, med_k, max_k)$, we define the five values as follows.

(1) $min_k = \min Set_{TI_k}$
(2) $mod_k =$ the most frequent value in Set_{TI_k}
(3) $ave_k =$ the average of the values in Set_{TI_k}
(4) $med_k =$ the median of the values in Set_{TI_k}
(5) $max_k = \max Set_{TI_k}$

Given a time interval $X_j = (min_j, mod_j, ave_j, med_j, max_j)(1 \leq j < n)$, if the equation $min_j = max_j$ holds, then the time interval between item i_j and item i_{j+1} is consistent; in particular, if the equation $min_j = max_j = 0$ holds, then these two items occurred at the same time.

Definition 6. *T-closed frequent sequential pattern A*

Given a T-sequential database D, let Σ be a set of T-frequent sequential patterns extracted from D and let A be a T-frequent sequential pattern of Σ. A is a **T-closed frequent sequential pattern** if Z, which is satisfying the following, does not exist in $\Sigma \backslash A$.

(1) If we let A' and Z' be O-patterns of A and Z, respectively, then $A' \subseteq Z'$.
(2) $Sup(A) \leq Sup(Z)$, where we define a support of the T-frequent sequential pattern as $Sup(A) \equiv | \{s \mid s \subseteq S, (sid, S) \in D$, where sid is the identifier of S in $D\} |$.
(3) If we let A and Z be $< a_1, T_1, a_2, T_2, \cdots, a_{n-1}, T_{n-1}, a_n >$, and $< z_1, T'_1, z_2, T'_2, \cdots, z_{m-1}, T'_{m-1}, z_m >$, respectively, then $j_1, j_2, \cdots, j_n$ exists and satisfies (1) $1 \leq j_1 < j_2 \cdots j_n \leq m$ and (2) $a_k = b_{j_k}, a_{k+1} = b_{j_{k+1}}$. Thus, for all $T_k = (min_k, mod_k, ave_k, med_k, max_k)$ and $T'_{j_k} = (min'_{j_k}, mod'_{j_k}, ave'_{j_k}, med'_{j_k}, max'_{j_k})$, equations (1) $min_k \geq min'_{j_k}$ and (2) $max_k \leq max'_{j_k}$ hold.

T-PrefixSpan outputs the set of T-frequent sequential patterns P with T-sequential database D and minimum support $MinSup$ as input.
More detailed definitions are given in [1].

3.2   Grouping the Clinical Pathways

We explained how to extract typical pathways without variants from EMR logs. In this section, we explain how to group the clinical pathways. That is, to detect a more practical variant from a combination of similar pathways. For example, two medical treatments in the variant have the same *Class*, *Description* and *Code* but not the same *Name*.

To achieve this, we group pathways. In each group, the number of medical treatments are all equal for each relative treatment day. The reference date of the relative treatment day is defined as the day on which the main medical treatment is performed. For example, the relative treatment day of the treatment done the day before the main medical treatment day is "–1" and one done on the day following the main medical treatment day is "1".

This date is determined from time intervals. The method in [1] defines time intervals that have five values, i.e., *minimum, most frequent, average, intermediate, maximum*, but we use only the most frequent value to calculate relative days. Because we think that typical sequential patterns are frequent patterns of EMR logs, relative days use the most frequent values.

3.3   Detecting Clinical Pathways Containing Variants

We have explained grouping the pathways in Sect. 3.2. In this section, we explain how to detect clinical pathways containing variants while treating time information from typical pathways without variants for each group.

First, we define the concepts required to introduce our method before we explain the algorithm.

Definition 7. *T-block B*

T-block B is the set of T-items with the same time of occurrence, which is denoted by $B = \{(i_1, t_1), (i_2, t_2), \cdots, (i_n, t_n) | t_1 = t_2 = \cdots = t_n\}$. Furthermore, let n be the number of elements of T-block B, and let t_B be the time at which these items occur, that is, $t_1 = t_2 = \cdots = t_n = t_B$.

Definition 8. *Variant pattern v*

Variant Pattern v is a sequence of T-blocks, which is denoted by $V =< (B_1, t_{B_1}), (B_2, t_{B_2}), \cdots, (B_n, t_{B_n}) >$. T-blocks that occur at the same time should be arranged in alphabetical order. Let n be the length of Variant Pattern V. When the number of elements of all the blocks is 1, the variant pattern is equal to the T-sequence.

We developed a method for detecting the pathway with variants from the pattern without variants. The algorithm is described in **Algorithm 1**. A variant pattern is detected by finding the difference between the clinical pathways in each group.

Algorithm 1. Detecting clinical pathways containing variants

Input: P' : the set of T-closed frequent sequential patterns
Output: V : the set of variant patterns
1: $v \leftarrow \{s \mid s \in S\}$
2: **for** $p \in P' \setminus v$ **do**
3:     $k, j = 1$
4:     **while** $k < length(v)$ and $j < length(p)$ **do**
5:         $B_v = \{B_i \in v\}$
6:         $B_s = \{B_j \in p\}$
7:         **if** $t_{B_v} == t_{B_p}$ **then**
8:             **for** $\{i \mid i \in B_v\}$ **do**
9:                 **if** i_{B_p} does not match any element of i_{B_v} **then**
10:                     Add (i_{B_p}, t_{B_p}) to B_v
11:                 **end if**
12:             **end for**
13:             $k = k + 1, j = j + 1$
14:         **else if** $t_{B_v} > t_{B_p}$ **then**
15:             Insert B_p just before B_v
16:             $k = k + 1, j = j + 1$
17:         **else**
18:             $k = k + 1$
19:         **end if**
20:     **end while**
21:     **while** $j < length(p)$ **do**
22:         Add $B_j, B_{j+1}, \cdots, B_{length(p)}$ to the end of v
23:     **end while**
24: **end for**
25: $V \leftarrow v$

3.4   Representing Variant Patterns

As a solution to the problem that cannot be expressed when the order cannot be uniquely determined, the method of representing the variants is defined by a nested structure with reference to the graphical notation [8].

Definition 9. *Nested branched sequence*

The items in the odd level of a list indicate a sequence pattern, while the items in the even level of the list a parallel pattern. $[L_{1,1}, L_{1,2}, ..., L_{1,n}]$ is a sequence pattern, $L_{1,i} = [L_{2,i,1}, L_{2,i,2}, ..., L_{2,i,m}]$ is a parallel pattern, and $L_{2,i,j} = [L_{3,i,j,1}, L_{3,i,j,2}, ..., L_{3,i,j,k}]$ is a sequential pattern, and so on.

For example, when five T-items(A,B,C,D,E) are written as $v = [[A], [B], [C, D], [E]]$, the variant pattern is shown in Fig. 1.

3.5   Visualizing Variant Patterns

The problem with the representation in the previous section is that if the structure becomes complicated, it becomes impossible to understand information

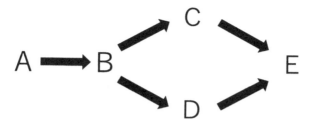

Fig. 1. Example of variant representation

intuitively, and the amount of information that can be read by each person is different. To provide the same information to anyone, visualization of data in an interactive graphical interface system is important.

3.6 Deriving the Set of Nodes and Edges from the Variant Pattern

For visualization, we must create sets of nodes and edges from the variant patterns. The algorithm is described in **Algorithm 2**.

Algorithm 2. Derivation of the set of nodes and edges

Input: Variant pattern v'
Output: the set of nodes of v' (V) and the set of edges of v (E)
  1: $k = 1$
  2: $count = 1$
  3: **for** $k < length(v)$ **do**
  4:      $start_{B_k} = count$
  5:      **for** $\{i \mid i \in B_k\}$ **do**
  6:         $i_{count} = i$
  7:         $V \leftarrow i_{count}$
  8:         **for** $k \geq 2$ and $start_{B_{k-1}} \leq j \leq end_{B_{k-1}}$ **do**
  9:            $E \leftarrow (i_j, i_{count})$
10:         **end for**
11:         $count = count + 1$
12:      **end for**
13:      $end_{B_k} = count - 1$
14:      $k = k + 1$
15: **end for**

The set of nodes contains all the T-items that exist in the variant pattern, the set of edges contains all edges from all T-items belonging to B_{k-1} to all T-items belonging to B_k. In the visualization, the graph is created by the set of nodes and edges.

4   Experiment

4.1   Experimental Data

We used target medical treatment data based on clinical pathways recorded from November 19, 1991, to October 4, 2015, in the EMRs at the Faculty of Medicine, University of Miyazaki Hospital. These medical data were acquired using an EMR system WATATUMI [9] employed by the Faculty of Medicine, University of Miyazaki Hospital. The total data size of the EMR system is 49 GB.

For personal information protection, the data that we used did not include information that could identify a patient uniquely. When we extracted the medical treatment data, we used anonymous patient IDs, which were impossible to associate with real people. The data we extracted from the EMRs to support medical treatments at the Faculty of Medicine, University of Miyazaki Hospital were described previously in [10] and they can be accessed at the website of the University of Miyazaki and the Research Ethics Review Committee of Tokyo Institute of Technology.

Our target data consisted of medical treatments based on two clinical pathways that were included in the EMRs: (1) *Transurethral Resection of a Bladder tumor (TUR-Bt)* and (2) *Endoscopic Submucosal Dissection (ESD)*. We chose these two clinical pathways because (1) TUR-Bt has a clinical pathway that is not clearly defined, whereas (2) ESD has clinical pathways that are relatively fixed.

In the experiment, we confirm that *variant pattern v* can be constructed For the two cases TUR-Bt and ESD.

The number of sequences, average length of the sequences and maximum length of the sequences for the two data sets (1) *TUR-Bt* and (2) *ESD* are shown in Table 1.

The reason for limiting the hospitalization period is to exclude exceptional pathways. For example, there is a pathway in which the treatment period should end in several days but the hospitalization period has exceeded one year.

Table 1. Target dataset

Dataset	(1) TUR-Bt	(2) ESD
Number of patients	242	46
Hospitalization period (min days/Max days)	4/11	8/11
Average treatments per patient	47.5	77.6
Maximum treatments per patient	118	136
Minimum treatments per patient	16	33

4.2 Results and Discussion

Confirming variant patterns: The numbers of typical clinical pathways by T-PrefixSpan are shown in Table 2 and the number of pathways constituting each variant pattern are shown in Tables 3 and 4.

Table 2. Numbers of typical clinical pathways by T-PrefixSpan

Dataset	(1) TUR-Bt		(2) ESD	
Threshold	0.5	0.6	0.5	0.6
Number of typical pathways	346	122	80	63
Average treatments per pathway	6.5	5.8	7.3	6.9
Maximum treatments per pathway	9	8	11	10
Minimum treatments per pathway	2	2	3	3

Table 3. Number of pathways constituting a variant pattern (threshold: 0.6)

Number of pathways constituting a variant pattern	1	2	3	4	5	6	7	8	9	10	11	12	Total
Number of variant patterns (TUR-Bt)	10	3	3	3	1	0	1	0	3	0	2	1	24
Number of variant patterns (ESD)	13	9	3	3	0	0	0	0	0	0	0	0	28

Table 4. Number of pathways constituting a variant pattern (threshold: 0.5)

Number of pathways constituting a variant pattern	1	2	3	4	5	6	7	8	9	Total
Number of variant patterns (TUR-Bt)	17	13	5	8	2	1	0	0	2	48
Number of variant patterns (ESD)	21	10	4	4	0	0	0	0	0	39

In the result of TUR-Bt, there are a few cases where a variant pattern is formed from many pathways. On the other hand, the result of ESD do not have such a case. This reason is that many small differences of medical treatments between pathways were detected because the clinical pathway of TUR-Bt is not fixed but there are few similar pathways because the clinical pathway of ESD is fixed.

As the threshold is lowered, total number of variant patterns increased in both cases, but the trend do not change, so we assume that there is no change in the trend even with a smaller threshold, for example, threshold is 0.1.

In all cases, it can be confirmed that variants can be constructed. This method is considered to be effective for pathways of any nature; in particular, where the pathway is not fixed, it is possible to detect variant patterns effectively. However, about half of the typical pathways that are not grouped are generated in any case. This indicates that the method of grouping can be improved.

A part of clinical pathways of TUR-Bt at threshold 0.1 for T-PrefixSpan are shown in Fig. 2, and variant patterns derived from Fig. 2 are shown in Fig. 3. The original character outputs are not colored, but we give the same color to the medical treatments of the same T-block in Fig. 2 and 3, for the readability.

The reason for selecting the low threshold is to display a variant pattern with many variants. Variant patterns at Fig. 3 is detected from five T-closed clinical pathways of Fig. 2, i.e., from patterns *a* to *e*. These pathways are expressed by nested branched sequence defined in Sect. 3.4.

```
<T-closed frequent sequential pattern : a>
···,[("order_type" : "Surgery" , "order_explain" : "TUR-Bt" , "order_name" : "" , "code" : "" , "day" : "0") ],
[("order_type" : "Injection" , "order_explain" : "Intravenous Injection" , "order_name" : "Lactec Injection" , "code" : "331" , "day" : "0") ],
[("order_type" : "Injection" , "order_explain" : "Intravenous Injection" , "order_name" : "Cefazolin Sodium for I.V. Infusion" , "code" : "613" , "day" : "0") ],
[("order_type" : "Pathological Diagnosis" , "order_explain" : "T ?M" , "order_name" : "" , "code" : "" , "day" : "0") ],
[("order_type" : "emergency test" , "order_explain" : "W B C" , "order_name" : "" , "code" : "" , "day" : "0") ],···

<T-closed frequent sequential pattern : b>
···,[("order_type" : "Surgery" , "order_explain" : "TUR-Bt" , "order_name" : "" , "code" : "" , "day" : "0") ],
[("order_type" : "Injection" , "order_explain" : "Intravenous Drip Injection" , "order_name" : "Veen-D Injection" , "code" : "331" , "day" : "0") ] ,
[("order_type" : "Injection" , "order_explain" : "Intravenous Injection" , "order_name" : "Lactec Injection" , "code" : "331" , "day" : "0") ],
[("order_type" : "Injection" , "order_explain" : "Intravenous Injection" , "order_name" : "Cefazolin Sodium for I.V. Infusion" , "code" : "613" , "day" : "0") ],
[("order_type" : "emergency test" , "order_explain" : "W B C" , "order_name" : "" , "code" : "" , "day" : "0") ],···

<T-closed frequent sequential pattern : c>
···,[("order_type" : "Surgery" , "order_explain" : "TUR-Bt" , "order_name" : "" , "code" : "" , "day" : "0") ],
[("order_type" : "Surgery Anesthesia" , "order_explain" : "closed-circuit anesthesia" , "order_name" : "" , "code" : "" , "day" : "0") ],
[("order_type" : "Injection" , "order_explain" : "Intravenous Injection" , "order_name" : "Lactec Injection" , "code" : "331" , "day" : "0") ],
[("order_type" : "Injection" , "order_explain" : "Intravenous Injection" , "order_name" : "Cefazolin Sodium for I.V. Infusion" , "code" : "613" , "day" : "0") ],
[("order_type" : "emergency test" , "order_explain" : "W B C" , "order_name" : "" , "code" : "" , "day" : "0") ],···

<T-closed frequent sequential pattern : d>
···,[("order_type" : "Surgery" , "order_explain" : "TUR-Bt" , "order_name" : "" , "code" : "" , "day" : "0") ],
[("order_type" : "Surgery Anesthesia" , "order_explain" : "closed-circuit anesthesia" , "order_name" : "" , "code" : "" , "day" : "0") ],
[("order_type" : "Injection" , "order_explain" : "Intravenous Injection" , "order_name" : "Lactec Injection" , "code" : "331" , "day" : "0") ],
[("order_type" : "Injection" , "order_explain" : "Intravenous Injection" , "order_name" : "Cefazolin Sodium for I.V. Infusion" , "code" : "613" , "day" : "0") ],
[("order_type" : "emergency test" , "order_explain" : "W B C" , "order_name" : "" , "code" : "" , "day" : "0") ],···

<T-closed frequent sequential pattern : e>
···,[("order_type" : "Surgery" , "order_explain" : "TUR-Bt" , "order_name" : "" , "code" : "" , "day" : "0") ],
[("order_type" : "Surgery Anesthesia" , "order_explain" : "closed-circuit anesthesia" , "order_name" : "" , "code" : "" , "day" : "0") ],
[("order_type" : "Injection" , "order_explain" : "Intravenous Injection" , "order_name" : "Lactec Injection" , "code" : "331" , "day" : "0") ],
[("order_type" : "Injection" , "order_explain" : "Intravenous Injection" , "order_name" : "Cefazolin Sodium for I.V. Infusion" , "code" : "613" , "day" : "0") ],
[("order_type" : "emergency test" , "order_explain" : "W B C" , "order_name" : "" , "code" : "" , "day" : "0") ],···
```

Fig. 2. Clinical pathways of TUR-Bt at thresold 0.1 for T-PresixSpan

Visualization: Even the colored outputs as Fig. 3, it is difficult for medical workers to grasp well the variant patterns from the character strings. Therefore, we try to visualize the variant patterns using an interactive graphical output tool *D3.js* [11]. We use circles and arrows between them to make easier to grasp the entire clinical flow, while make detailed information be given interactively by user clicks.

···[{'order_type' : 'Prescription', 'order_explain' : 'external medicine' , 'order_name' : 'glycerin enema 「OHTA」 ' , 'code' : '235' , 'day' : '0'}],

[{'order_type' : 'Surgery' , 'order_explain' : 'TUR-Bt' , 'order_name' : '' , 'code' : '' , 'day' : '0'}],

[{'order_type' : 'Injection' , 'order_explain' : 'Intravenous Injection' , 'order_name' : 'Lactec Injection' , 'code' : '331' , 'day' : '0'} ,

{'order_type' : 'Injection' , 'order_explain' : 'Intravenous Drip Injection' , 'order_name' : 'Veen-D Injection' , 'code' : '331' , 'day' : '0'} ,

{'order_type' : 'Surgery Anesthesia' , 'order_explain' : 'closed-circuit anesthesia' , 'order_name' : '' , 'code' : '' , 'day' : '0'}],

[{'order_type' : 'Injection' , 'order_explain' : 'Intravenous Injection' , 'order_name' : 'Cefazolin Sodium for I.V. Infusion' , 'code' : '613' , 'day' : '0'} ,

{'order_type' : 'Injection' , 'order_explain' : 'Intravenous Injection' , 'order_name' : 'Lactec Injection' , 'code' : '331' , 'day' : '0'}],

[{'order_type' : 'Pathological Diagnosis' , 'order_explain' : 'T?M' , 'order_name' : '' , 'code' : '' , 'day' : '0'} ,

{'order_type' : 'Injection' , 'order_explain' : 'Intravenous Injection' , 'order_name' : 'Cefazolin Sodium for I.V. Infusion' , 'code' : '613' , 'day' : '0'}],

[{'order_type' : 'emergency test' , 'order_explain' : 'WBC' , 'order_name' : '' , 'code' : '' , 'day' : '0'}],···

Fig. 3. Variant patterns derived from Fig. 2

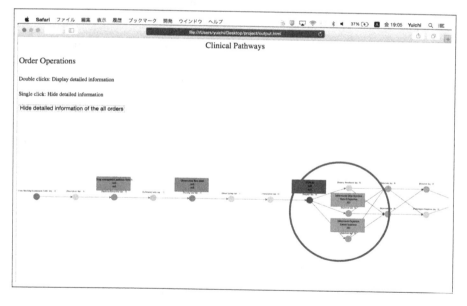

Fig. 4. Visualization example (all) (Color figure online)

For example, the variant patterns of TUR-Bt in Fig. 3 are visualized in Fig. 4. Figure 5 magnifies a part of variants indicated by the red circle in Fig. 4. Over the circle node, *Type* and *Treatment day* are always displayed because they are necessary information for grasping the entire pathway. We make the same color

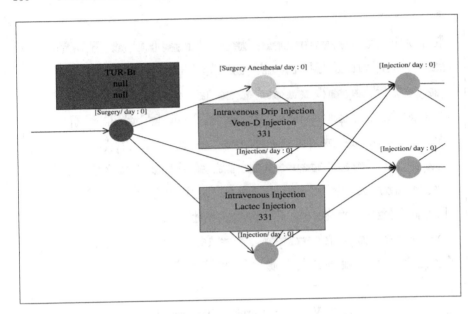

Fig. 5. Visualization example (part)

node have the same *Type* of medical treatment. When the mouse click a circle node, detailed information, i.e., *Explain*, *Code*, and *Name* are displayed in the square node. This square node with the detailed information is a toggle between hide and show to focus on important portions in the pathway.

In this example, two pink nodes match with *Type* injection, and *code* 331. This code indicates a blood substitute agent. However, the two values of *Name* are different (Veen-D Injection and Lactec Injection). By visualizing, these differences are easier to understand than character based results.

5   Conclusion

In this study, we have proposed a method for detecting variants in clinical pathways with treatment time information from EMR logs. We combine typical clinical pathways to find the differences between similar pathways. We then develop an interactive graphical interface system for visualizing the variants in clinical pathways. The graphical output helps to grasp outlines of medical treatment flows and to focus on the detailed information of user specified portions.

We confirmed the proposed method using actual EMR logs from a university hospital. The results show that variant patterns are detected for clinical pathways for diseases. Comparing the characteristics of diseases, the proposed method was especially effective where the pathway was not fixed. On the other hand, only about half of the variant patterns were detected. We plan to improve the method for grouping pathways.

It is currently unclear which action is more frequent in variants. We also plan to improve the method to provide the information about variant frequency. Additionally, new medical treatment paths that do not exist in extracting typical clinical pathways is detected in detection of variants, which is another future work. From the viewpoint of support for medical workers, it is important to find the causes of variants. We continue this approach to collaborate with medical workers using clinical pathways.

References

1. Uragaki., K., Hosaka., T., Arahori., Y., Kushima., M., Yamazaki., T., Araki., K., Yokota., H.: Sequential pattern mining on electronic medical records with handling time intervals and the efficacy of medicines. In: First IEEE Workshop on ICT Solutions for Health. Proceedings of 21st IEEE International Symposium on Computers and Communications, pp. 20–25 (2016)
2. Wakamiya, S., Yamauchi, K.: What are the standard functions of electronic clinical pathways? Int. J. Med. Inform. **78**, 543–550 (2009)
3. Hirano., S., Tsumoto., S.: Clustering of order sequences based on the typicalness index for finding clinical pathway candidates. In: IEEE International Conference on Data Mining ICDM Workshops (2013)
4. Agrawal., R., Srikant., R.: Fast algorithms for mining association rules in large databases. In: Proceedings of the 20th International Conference on Very Large Data Bases, pp. 487–499 (1994)
5. Chen, Y.-L., Chiang, M.-C., Ko, M.-T.: Discovering time-interval sequential patterns in sequence databases. Expert Syst. Appl. **25**, 343–354 (2003)
6. Pei., J., Han., J., Mortazavi-Asl., B., Pinto., H., Chen., Q., Dayal., U., Hsu., M-C.: PrefixSpan: mining sequential patterns efficiently by prefix-projected pattern growth. In: Proceedings of 2001 International Conference on Data Engineering, pp. 215–224 (2001)
7. Mannila, H., Toivonen, H., Verkamo, A.I.: Discovery of frequent episodes in event sequences. Data Min. Knowl. Disc. **1**, 259–289 (1997)
8. Achar., A., Laxman., S., Raajay., V., Sastry., P,S.: Discovering general partial orders from event streams. Technical report. arXiv:0902.1227v2 [cs.AI]. http://arxiv.org
9. Denshi Karte System WATATUMI (EMRs "WATATUMI"). http://www.corecreate.com/02_01_izanami.html
10. Miyazaki Daigaku Igaku Bu Fuzoku Byouin Iryo Jyoho Bu (Medical Informatics Division, Faculty of Medicine, University of Miyazaki Hospital). http://www.med.miyazaki-u.ac.jp/home/jyoho/
11. D3.js. https://d3js.org

Umedicine: A System for Clinical Practice Support and Data Analysis

Nuno F. Lages[1], Bernardo Caetano[1], Manuel J. Fonseca[2], João D. Pereira[1],
Helena Galhardas[1(✉)], and Rui Farinha[3]

[1] INESC-ID and Instituto Superior Técnico,
Universidade de Lisboa, Lisboa, Portugal
{nuno.lages,bernardo.silva.caetano,
helena.galhardas}@tecnico.ulisboa.pt, joao@inesc-id.pt
[2] LaSIGE, Faculdade de Ciências, Universidade de Lisboa, Lisboa, Portugal
mjfonseca@ciencias.ulisboa.pt
[3] Hospital de São José, Lisboa, Portugal
ruifarinhaurologia@gmail.com

Abstract. Recording patient clinical data in a comprehensive and easy way is very important for health care providers. However, and although there are information systems to facilitate the storage and access to patient data, many records are still in paper. Even when data is stored electronically, systems often are complex to use and do not provide means to gather statistical information about a population of patients, thus limiting the usefulness of the data. Physicians often give up searching for relevant information to support their medical decisions because the task is too time-consuming. This paper proposes Umedicine, a web-based software application in Portuguese that addresses current limitations of clinical information systems. Umedicine is an application for physicians, patients and administrative staff that keeps clinical data (e.g., symptoms, clinical examination results, and treatments prescribed) up to date on a database in a structured way. It also provides easy and quick access to a large amount of clinical data collected over time. Furthermore, Umedicine supports the application of a particular clustering algorithm and a visualization module for analyzing patient time-series data, to identify evolution patterns. Preliminary user tests revealed promising results, showing that users were able to identify the evolution of groups of patients over time and their common characteristics.

Keywords: Clinical information system · Data analysis · Clustering · Data visualization

1 Introduction

For health care providers, recording every patient's clinical information comprehensively is of paramount importance. However, for many years this information

© Springer International Publishing AG 2017
E. Begoli et al. (Eds.): DMAH 2017, LNCS 10494, pp. 102–120, 2017.
DOI: 10.1007/978-3-319-67186-4_9

has been recorded in paper and kept in large dedicated archives, making it difficult to retrieve relevant past clinical information quickly and effectively when required for patient care or for scientific research purposes.

With the widespread use of computers, new software tools were developed for clinical staff to record information about patients. This easy access to clinical data promises a significant impact in clinical practice. In particular, continuous patient monitoring can be ubiquitous, enabling fast response by clinical staff and quick situation assessment by physicians; clinical research can benefit from much larger, easier-to-access data sources, that will accelerate the discovery of new medical knowledge; and health care managers will be able to make more informed decisions regarding institutional policies.

Currently, existing medical information systems often are too complex to use and provide free-form fields for collecting patient's medical information, which is not stored as structured data. For this reason, many patient records are still in paper and, in the cases where the information is in the system, it is not possible to gather statistical information about a population of patients, thus limiting the usefulness of the data. As a consequence, physicians often give up searching for relevant information to support medical decisions because the task is too time-consuming [2,10].

Hence, there still is a need of tools to enable integration and analysis of clinical data in an effective manner [10,11] and to make it useful in everyday practice. Physicians should have means to explore existing patient data quickly to support their diagnosis, treatment decisions and research.

In this paper, we propose *Umedicine*, a new clinical information system that addresses the current limitations of this kind of systems. Umedicine is a web-based application with an appealing and easy-to-use user interface to be used by physicians and administrative staff in hospitals/clinics and patients at home. Through it, physicians and patients can enter clinical data into a platform that provides easy, quick and always-on access to a large amount of clinical data. Clinical data is persistently kept in an electronic and structured format in a database, thus enabling data analysis to extract interesting patterns and the application of visualization techniques to show these patterns to physicians. To the best of our knowledge, our application is the only one that supports: (i) filling in standard international diagnosis-support questionnaires, (ii) collecting and storing patient data (personal data and clinical data such as symptoms, clinical examination results, diagnosed pathologies and prescribed treatments) in a structured way that is appropriate to each medical specialty, and (iii) applying data analysis algorithms, all integrated in one software platform. In particular, we demonstrate the use of a clustering algorithm and a visualization mechanism for analysis of patient time-series data.

Umedicine is designed to support any medical specialty, and in this paper we used the Urology specialty to demonstrate the use of the application. We performed a preliminary evaluation of the Umedicine user interface with users, asking them to perform several task scenarios. Users also filled in a satisfaction questionnaire to assess Umedicine qualitatively. The results were promising, as users were able to complete with success all the tasks and were mostly satisfied with the user interface provided.

2   Architecture and Implementation

Umedicine has a client-server style architecture with three main components (see Fig. 1): (i) a front-end available to the user through a web browser client, (ii) a back-end server, and (iii) a relational database. Users interact with the system through a web browser, which submits HTTP requests to the server. The server returns an HTML page. Data may be asynchronously requested from the server through Asynchronous JavaScript and XML (Ajax) requests made by the browser. To respond to Ajax requests, the server queries a relational database and returns the data to the client in the form of JSON objects. The use of Ajax allows to exchange only new necessary data to fill the page, minimizing data transfers from the server and improving user waiting time. The server is developed in Java with extensive use of Spring Framework[1]. Web pages are generated with the JavaServer Pages (JSP) and Apache Tiles[2] technologies.

Fig. 1. Architecture of the Umedicine system

Umedicine supports four types of users: non-administrator physicians, administrator physicians, clerks and patients. Non-administrator physicians can add new patients to the system and search, read and modify patient personal and medical information. In addition to having access to all the functionalities available to non-administrator physicians, administrator physicians can add new clerks and physicians (both administrator and non-administrator) to the system. Patients can view and modify their own personal information, view their own examination results and history and fill medical questionnaires for diagnosis support. Clerks can add new patients to the system and add limited patient personal information (name, birth date, contact and profession). The application is implemented in a way that ensures that each type of user can only use the functionalities and have access to information as described above. Data confidentiality is guaranteed in communications between clients and server by use of Transport Layer Security (TLS) encryption.

---

[1] https://spring.io/.
[2] https://tiles.apache.org/.

2.1 Front-End

Umedicine is implemented as a web application, providing responsive user interfaces for the different types of user, and can be used both on mobile and desktop environments. In this section, we focus on the user interfaces offered for physicians and patients.

When the physician selects a patient, the patient's information page is shown, as illustrated in Fig. 2. It is organized in six parts:

- **Personal information** (*Informação Pessoal*): area where physicians can view and modify information such as name, birth date, contact and habits of the patient;
- **Disease** (*Doença*): area where physicians can select the medical condition that applies to the selected patient and view or modify information about symptoms; following the development team's physician advice, this area also includes the rectal examination information; users can view symptom and rectal examination histories and add new symptoms and examination results;
- **Treatment** (*Tratamento*): this area displays information about ongoing treatments and provides means to view treatment history and add new treatments;
- **Questionnaires** (*Questionários*): this area provides access to three internationally-accepted diagnosis-support questionnaires that a patient can answer at home or during a medical appointment; these questionnaires are used to compute scores that represent the severity of a patient's condition; Umedicine currently supports three relevant standard questionnaire-based scores: State Self Esteem Scale (IIEF), International Prostate Symptom Score (IPSS) and State Self Esteem Scale (SSES);
- **Diagnosis support examinations** (*Exames Auxiliares de Diagnóstico*): this area shows the most recent results for several kinds of medical examinations and laboratory tests; the user can add new results and access the patient's examination and test histories;
- **Notes** (*Observações*): physicians can write a textual note about the patient's condition on the day of the medical appointment and view notes from previous appointments.

The user interface available to patients is more limited (Fig. 3). It contains an area on the left where they can view their current photo and alerts requiring attention. The middle area shows their current personal information. On the right, patients can view information about examinations, current medications and past surgeries. At the bottom, patients can choose to edit their personal information or to answer a diagnosis support questionnaire. Note that these tasks can be performed away from the clinic, saving time during medical appointments and give more privacy to patients.

2.2 Back-End Server

The back-end server replies to the requests sent by the browsers, performs operations on the data and stores or retrieves data as necessary from the database.

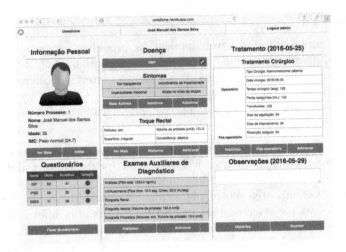

Fig. 2. A patient's information page as viewed by a physician using Umedicine

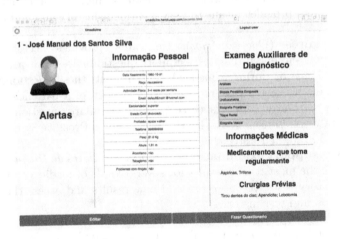

Fig. 3. A patient's information page as viewed by the patient using Umedicine

It is responsible for ensuring that each user has access to, and only to, the data they need, and for loading the data or web pages requested by the users. The server is implemented as a Java application with extensive use of the Spring Framework ecosystem[3]. It follows the Model-View-Controler (MVC) design pattern and is composed of three main layers — from top to bottom: controllers, services and Data Access Objects (DAOs) (Fig. 4).

Controllers are the components responsible for the interaction with the clients. They receive client requests, invoke the appropriate business logic methods and send a response if required. Controller code (as well as service and DAO

---

[3] https://spring.io.

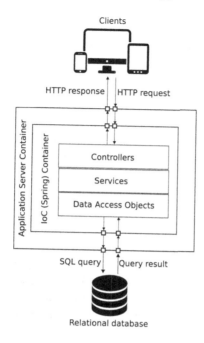

Fig. 4. Umedicine's server component architecture

code) is, as much as possible, organized to maximize separation of concerns. In other words, a controller is written to support a specific set of coherent functionalities. For example, the code responsible for handling requests related to user account creation is supplied by methods of the class *UserController*; requests from pages where users fill in questionnaires are handled by the *Questionnaire-Controller* class; and so on.

Any business logic-related processing of the data retrieved from the database or sent by the clients takes place in the *Services* layer. Service methods are invoked by controllers (or other services) and return data that controllers use to respond to the requests issued by the client. Service code is also organized according to functionalities. *UserService*, for instance, contains the code responsible for user account creation, performing duties such as invoking the Data Access Object (DAO) methods that store user data in the database (after performing any required data processing) or sending emails with new passwords to new users. In a similar fashion, there is a *QuestionnaireService* class that handles questionnaire data, and so on. Services also organize data retrieved from the database via DAOs into objects suitable to be used by the controllers, which may return data to the browser.

The bottom layer, DAOs, contains the code that enables the interaction between the server and the database. This code includes the SQL queries that insert new rows, update existing rows and retrieve data from the database tables.

Information retrieved from the database is converted to Java objects and passed on to the upper service layer for further processing.

2.3 Relational Database

Umedicine's architecture encloses a relational database to store three kinds of data persistently: (i) clinical data inserted by patients or physicians; (ii) user authentication data; and (iii) metadata concerning a specific medical field. The current version of the application has a database that is set up for Urology but its schema is designed to be adaptable to other medical specialties. For this end, the metadata that corresponds to domain knowledge – drug names, disease names, symptoms, medical examinations and diagnosis-support questionnaires – is stored in the database rather than in the back-end server and is accessed when the application loads. With this approach, different domain knowledge from other medical specialties can be substituted into the database and plugged into the server with minimal modification of server and client code (ideally no modification will be needed whatsoever). This approach also enables switching easily from metadata in Portuguese to metadata in another language.

To illustrate Umedicine's database model, we present the subset of the database relational schema that models medical examinations. Primary keys of relations are underlined and foreign keys are specified by FK:

Clinical metadata:

ExaminationType (<u>typeName, subTypeName</u>)
Parameter (<u>paramName</u>, paramUnit, paramType)
ExaminationParameter (<u>typeName, subTypeName, paramName</u>)
    typeName, subTypeName: FK (ExaminationType)
    paramName: FK (Parameter)

Clinical data:

Patient (<u>pNumber</u>)
PerformedExamination (<u>pExamID</u>, pNumber, typeName, subTypeName, date)
    pNumber: FK (Patient)
    typeName, subTypeName: FK (ExaminationType)
PerformedExaminationValue (<u>pExamID, paramName</u>, value)
    pExamID: FK (PerformedExamination)
    paramName: FK (Parameter)

There are two types of data modeled in this relational schema: (i) clinical metadata and (ii) clinical data. ExaminationType, Parameter and ExaminationParameter are relations that model metadata. This metadata is accessed when the application starts and is used to determine the information that is shown in the user interface. The ExaminationType table stores all the possible types and subtypes of medical examinations. The tuples: $<Urofluxometry, - >$ and $<BloodTests, Liverfunction>$ are examples of records stored in this table.

Volume (mL)

Fig. 5. Parameter Volume of type FLOAT and unit milliliter (mL)

Urofloxometry is a medical examination by itself while there are several kinds of blood tests so the subtype field is required. The Parameter table stores the existing parameters, their measurement units and expected types, i.e., String, Integer, Float or enumerated (in the case the type is ENUM, there is another table not represented here that stores the possible values). The tuples: $<Creatinine, mg/dl, float>$ and $<Volume, ml, float>$ are examples of records of the Parameter table. The table ExaminationParameter stores the correspondence between an examination and its parameters. In accordance with the previous examples, the following tuples: $<Urofluxometry, -, Volume>$ and $<BloodTests, Liver function, Creatinine>$ are examples of tuples of the ExaminationParameter table.

The tables Patient, PerformedExamination and PerformedExaminationValue store clinical data collected by the application. In particular, Patient stores data about patients ($<P1>$ and $<P2>$ are examples of tuples); PerformedExamination stores the examinations performed by each patient ($<PE1, P1, BloodTests, Liver function, 27/6/2016>$ and $<PE2, P2, Urofluxometry, -, 28/7/2016>$ are examples of tuples); and PerformedExaminationValue stores data about the filled parameters of the examination performed by a patient (examples of tuples are: $<PE1, Creatinine, 1.02>$ and $<PE2, Volume, 157>$).

Figure 5 shows the application screen shown for the examination parameter Volume, which is of type float and is measured in milliliters. The metadata stored in the database table Parameter, in particular, supplies the (meta)data to be shown in the forms presented in the front-end user interface. The clinical data that will be filled in by the users is then stored in the database tables that store clinical data and is further available for future visualization and analysis.

3   Clinical Time-Series Data Clustering and Visualization

A major advantage of medical information systems is the fact that they are capable of storing data from a large population of patients, hence providing a large source of data for discovering trends in disease progression that might be difficult to uncover by physicians in their daily clinical practice. In order to make such datasets useful, the medical information system needs to provide means to find groups of similar patients with basis on their personal and medical history and enable the exploration of the characteristics of the patients in these groups. We propose an approach to address this requirement, using an off-the-shelf state-of-the-art clustering algorithm for time series to find groups of patients with similar variation of relevant clinical parameters. We complemented the data analysis performed by the clustering algorithm with a visualization module to enable physicians to explore the results for each group of patients discovered by the algorithm. This approach can be applied to any medical parameter that varies over time.

3.1 Creating Time-Series Data Clusters

Clustering algorithms are widely used in biological research, among many other fields, for a variety of tasks [9]. The use of classical clustering techniques such as k-means has important disadvantages when used with biological or clinical data:

1. These algorithms tend to find clusters of similar size, hence they may not find interesting, relatively small clusters.
2. Each element (patient) is assigned to one and only one cluster, while it may display behavior similar to more than one (albeit at different intervals in time), or to none.
3. These algorithms compute clusters using all dimensions (which, in time-series data, translate to time points) at once, hence they may not cluster together time series that are similar in all but one or two time points.

Biclustering algorithms are a family of clustering algorithms that overcome these disadvantages. Due to the similar size of genomic and medical datasets, we expect that biclustering's benefits apply to medical time series as well. Biclustering algorithms find groups of patients with basis on a subset of the time points instead of using all time points at once. In other words, they produce a local model instead of a global model [5]. The input data of biclustering algorithms is represented as a data matrix, where each row represents the evolution of a given medical parameter for a patient, each column represents a time point and each matrix element holds a value for a medical parameter. Biclustering algorithms find clusters of rows in the input data matrix that are similar across a subset of the columns. The complexity of biclustering algorithm depends on the applied criteria of similarity between rows, but most formulations are NP-hard [5,7]. Hence, most biclustering algorithms resort to heuristics without guaranteeing optimal solutions, or have prohibitive running times. The problem becomes tractable, however, in cases where the measured medical parameter is discrete and the search for similar series is restricted to contiguous time points [5]. The Contiguous Column Coherent Biclustering CCC-Biclustering algorithm [5] overcomes these restrictions to find clusters in time-series data in time linear in the size of the input data matrix. However, many medical parameters of interest are not discrete. In the original paper, the authors are concerned with changes in gene expression, which are measured in a continuous positive scale with fixed zero minimum. Hence, the authors apply a discretization strategy to represent the variation of the gene expression in three levels: significant increase, significant decrease and no change of gene expression. This strategy reflects their goal of finding groups of genes with similar expression variation patterns. For other problems, care must be taken to choose an appropriate discretization approach as well.

Due to its performance, we decided to use the CCC-Biclustering algorithm in Umedicine. We illustrate the application of CCC-Biclustering in Umedicine with International Prostate Symptom Score IPSS time-series data, but this algorithm can be applied to any clinical parameter that is measured at different time instants. IPSS is a score calculated from a standardized questionnaire given to certain Urology patients, such as patients that suffer from Benign Prostatic Hyperplasia (BPH). Physicians request that BPH patients fill in the questionnaire several times during the treatment, obtaining scores that reflect the evolution of patients over time. IPSS has a uniform scale of integer values between 0 (best-case scenario for the patient) and 35 (worst-case scenario for the patient). It is, thus, discrete. However, imposing such a fine discrete scale of 36 values would rarely cluster patients together: only when they had a sequence of exactly equal scores would the algorithm consider them part of the same cluster. For this reason, it makes sense to use a coarser discretized scale: for example, a scale with 6 discretization levels would enable for a difference of up to 6 points in IPSS while still being able to capture increasing or decreasing trends, as variations of the IPSS value would often result in a change to a different level.

The CCC-Biclustering library made available by its authors also computes statistics for each cluster that it finds. One of the most important of these statistics is a p-value that reflects how similar the patients in the same group are among themselves. This p-value is calculated with basis on a hypothesis test in which the null hypothesis assumes that a cluster, with its size and patient data, was randomly generated. The lower the value of the p-value, the smaller is the probability of finding this group of patients under this null hypothesis. Hence, patients in groups with lower p-values can be expected to be more similar.

3.2 Visualization of Clinical Time-Series Data Clusters

We developed an interactive visualization mechanism and incorporated it into the Umedicine application, to enable the exploration of time-series data as analyzed by the CCC-Biclustering algorithm described in Sect. 3.1. The visualization is composed of a matrix of line charts (illustrated in Fig. 6), where each chart corresponds to a cluster of patients computed by the CCC-Biclustering algorithm. Each chart also indicates the number of patients in the corresponding group.

Users can set six different parameters to modify the visualization. Some of these parameters correspond to parameters of the CCC-Biclustering algorithm but were renamed for a more intuitive use by users with little knowledge of statistics. In the same order as in Fig. 6, they are:

1. **Allowed similarity between groups** (*Semelhança permitida entre grupos*): this parameter corresponds to the maximum overlap between clusters computed by the CCC-Biclustering algorithm. In other words, it defines the maximum percentage of patients in a group that can also be part of another group. The user can choose five levels of similarity: very small, small, intermediate, large and very large, corresponding to 1%, 5%, 10%, 25% and 50% maximum overlap.

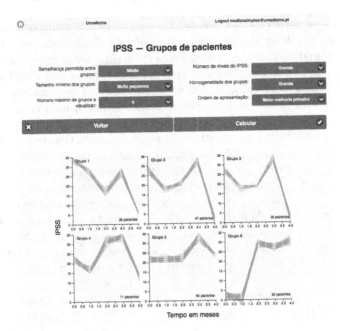

Fig. 6. Visualization of time-series clustering in Umedicine for the IPSS parameter

2. **Number of IPSS levels** (*Número de níveis do IPSS*): corresponds to the number of intervals/symbols chosen to discretize the IPSS scale. The larger the number of IPSS levels, the narrower they are. Hence, a larger number of levels leads to clusters where patients are more similar to each other. The user has five options: very small, small, intermediate, large and very large, corresponding to 3, 6, 9, 12 and 18 levels of discretization.

3. **Minimum group size** (*Tamanho mínimo dos grupos*): the minimum number of patients in each cluster to be displayed. Charts with a number of patients inferior to the chosen number are not shown. The user has five options: very small, small, intermediate, large and very large. The 'very small' option sets the minimum number of patients per cluster/chart to two. The other options are computed as a function of the total number of patients in the dataset and correspond to 10%, 20%, 40% and 60% of the total number of patients, respectively.

4. **Group homogeneity** (*Homogeneidade dos grupos*): the larger this parameter, the smaller the maximum p-value of the clusters — as explained in Sect. 3.1; clusters whose p-values exceed this value are not shown. The user chooses among the options 'very small', 'small', 'intermediate', 'large' and 'very large', which correspond to p-values 0.25, 0.125, 0.083, 0.0625 and 0.05.

5. **Maximum number of clusters**/charts to show (*Número máximo de grupos a visualizar*): it takes into account the chosen presentation order of charts to filter out exceeding clusters.

6. **Presentation order** (*Ordem de apresentação*): users can choose between displaying charts ordered by decreasing number of patients per cluster, by decreasing overall change in IPSS or by increasing overall change in IPSS. The overall change in IPSS here is the simple difference between IPSS at the final time point and the IPSS at the initial time point.

After setting these parameters as desired, the user generates the visualization by clicking/tapping the 'Calculate' (*Calcular*) button shown in Fig. 6.

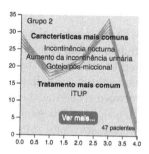

Fig. 7. Example of information displayed over a chart (from Fig. 6)

Each chart can be selected with a mouse click or finger tap. This action overlays information about the corresponding group of patients, as illustrated in Fig. 7. The information shown includes the three most prevalent medical parameters in the patients belonging to the selected group (*Características mais comuns*) and the treatment given to most patients (*Tratamento mais comum*). Pressing the button at the bottom of the chart (*Ver mais...*) shows a table with detailed information about the selected group of patients (Fig. 8). This information includes the percentages of patients that were subject to the most common treatments for this group, statistical information about weight, prostate volume and age of these patients and, for each symptom and medical characteristic, the percentage of patients of the cluster that has this symptom or characteristic. This last part is sorted from highest to lowest.

Informações sobre o grupo 2

Característica	Informação
Tratamentos	Alfa-bloqueante: 19.4%, Adenomectomia: 16.7%; Nenhum: 13.9%;
Peso	Média: 91.8; Mediana: 91.8; Desvio padrão: 5.9
Volume da Próstata	Média: 31.5, Mediana: 31.8; Desvio padrão: 2.3
Idade	Média: 38.7; Mediana: 39; Desvio padrão: 6.2
Nódulos na próstata	97.2 %
Diminuição de força e calibre do jacto	94.4 %
Noctúria	91.7 %

Fig. 8. Example of information displayed when a user clicks the button 'Ver mais' (meaning 'get more information') on one of the charts shown in Fig. 7

4 Experimental Validation

To validate our solution, we performed: (i) performance tests to analyze the behavior of the clustering algorithm according to the number of patients and the number of data points over time; and (ii) usability tests to evaluate the technique for visualizing and inspecting the clusters of patients produced by the clustering algorithm.

4.1 Performance Evaluation

To evaluate the performance of the clustering algorithm with medical time-series data, we varied two input parameters: number of patients and number of time points per patient. We measured the time spent to analyze the data (clustering) and to display the results of the analysis (visualization). For the analysis (server side), we measured the time between receiving the request from the client up to the response from the server. Hence, it includes reading the data, computing the clusters, computing the related statistics and organizing data for presentation. Concerning the performance of displaying the results (client side), we measured the time between receiving the cluster data from the server and displaying the charts on the screen.

Experimental Setup. To perform the evaluation of the clustering algorithm we had to generate synthetic data, because at this phase of the project we do not yet have real data from patients. We generated a synthetic dataset to be representative of a collection of patients that answered the IPSS questionnaire at several points in time after beginning a treatment[4]. In addition to the IPSS scores we also generated data for various parameters relevant to the BPH pathology (e.g. weight, prostate volume).

We performed two experiments to test the CCC-Biclustering. In the first experiment, we varied the number of patients for a fixed number of time points (5), and in the second, we varied the number of time points for a fixed number of patients (1,000). The fixed values for time points and patients were chosen to generate datasets with data close to reality. The experiments were performed on a 2009 MacBook Pro laptop computer with 8 GB of 1066 MHz DDR3 memory, a 2.66 GHz Intel Core 2 Duo processor and a SATA hard disk drive.

Results and Discussion. The results obtained for both experiments are shown in Fig. 9. The left panel shows the behavior of our solution when we vary the number of patients and the right panel shows the behavior of our solution when we vary the number of time points per patient.

According to the CCC-Biclustering authors, the algorithm has a worst-case complexity linear in the size of all input parameters. Our results comply with this

---

[4] We are aware that synthetic data may not have the same behaviour as real data, but our goal was to identify groups of patients.

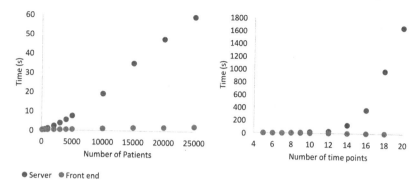

Fig. 9. Performance measurements for the CCC-Biclustering algorithm and the cluster visualization mechanism. Left panel: results for different numbers of patients for a fixed number of 5 time points per patient. Right panel: results for different numbers of time points per patient, for a fixed number of 1,000 patients.

expectation when we vary the number of patients (Fig. 9-left), but not when we vary the number of time points (Fig. 9-right). We speculate that this deviation from linearity is caused by data pre- and/or post-processing steps, such as the computation of statistics. However, further investigation is necessary to explain this result.

Although the time to compute clusters for a large number of time points is relatively high, for realistic scenarios (less than 14 time points, corresponding to 14 appointments) the computation on the server side takes less than three minutes, making it possible to be used in an appointment scenario, where physicians can perform the analysis while attending the patient. The results only suffer a degradation of performance (with waiting times higher than 3 min) for more than 14 time points per patient, an unlikely scenario.

The performance measured for increasing number of patients is also encouraging: a matrix with 25,000 patients and the realistic column size of 5 time points could be analyzed in one minute. Assuming a linear trend, this means that data from hundreds of thousands of patients could be analyzed in a few minutes.

The processing of the visualization in the front end, in turn, added a comparatively small waiting time, typically under half a second, which does not represent a performance concern.

4.2   Usability Tests

To complement the performance tests, we performed a usability test with users, to evaluate the visualization and inspection technique used to present the groups of patients and their main characteristics.

Experimental Setup. We recruited ten volunteers to participate in the tests, half of them male. Their ages were between 24 and 34 years old and all of them had a university degree. The test consisted of using the application to perform six tasks (e.g. identify the group of patients that improved their condition, the main characteristics of this group of patients, etc.). After completing the tasks, we asked users to answer a satisfaction questionnaire, composed of seven questions to rate the usage of our visualization mechanism and its characteristics (e.g. color scheme, presentation of information), and three open-answer questions inquiring about the main difficulties, best features and suggested modifications to the visualization.

Results and Discussion. Table 1 summarizes information about the users and how they rated the visualization (from 1 to 5, 5 being the best) on several topics.

Table 1. Usability test results

User	1	2	3	4	5	6	7	8	9	10	Average	Standard error
Sex	F	M	F	M	F	M	M	M	F	F	NA	NA
Age	26	24	33	33	33	26	33	33	34	31	30.6	3.7
YUC	15	16	16	18	20	27	22	20	24	26	20.4	4.3
Q1	5	4	4	4	5	5	4	4	5	5	4.5	0.5
Q2	4	3	3	4	4	4	4	4	4	4	3.8	0.4
Q3	5	5	4	5	5	5	5	5	3	5	4.7	0.7
Q4	4	5	4	5	5	4	4	5	4	5	4.5	0.5
Q5	5	5	4	4	5	5	4	5	4	5	4.6	0.5
Q6	4	5	4	4	5	3	3	4	4	5	4.1	0.7
Q7	5	5	5	4	5	5	4	4	5	5	4.7	0.5

YUC: How long have you been using computers, in years?
Q1: The color scheme is appropriate to visualize the data
Q2: The options given to control patient group generation are easy to understand
Q3: The user can understand when a patient improves or worsens
Q4: The user can understand that patients in the same group have similar IPSS along time
Q5: The user gets a good understanding of the characteristics of patients in the same group
Q6: It is easy to understand why patients are grouped together
Q7: The data displayed in the visualization table complement well the information shown in the corresponding chart

Overall, the results shown in Table 1 indicate that the users were pleased with their experience regarding the issues in questions Q1 to Q7. The lack of clarity of the meaning of the visualization parameters was the main complaint during the test. However, eight of the ten users recognized that once the meaning of the options was understood, they could use them and interpret how they impacted the observed results. The experiment showed that the presentation of the options should be revised but, in general, users were able to complete the tasks successfully.

In the last part of the test, we asked users to describe what, in their opinion, were their main difficulties, the most positive aspects of the visualization and what they would like to see changed. Most users complained explicitly about the unintuitive options provided. Several users suggested to have the button that closes the table staying visible while the user scrolls up or down to inspect the table. Users tended to complimented the look and organization of the visualization page, with two of them mentioning the 'clean' look as a very positive aspect. Five users pointed the identification of common characteristics of the groups (Fig. 7) and the table with group statistics (Fig. 8) as the most positive aspects of the visualization. Three users indicated that they liked the interaction experience, especially its responsiveness. Other positive comments mentioned that it was easy to understand when patients improve or worsen.

5   Related Work

This section provides an overview of: (i) the most relevant commercial medical software applications that offer functionalities similar to the ones offered by Umedicine; and (ii) research works focusing on some of Umedicine's features.

There is a vast amount of medical software applications available in the market, providing support to clinical practice (in particular clinical appointments). The most important features offered are: (i) health center management aiming at managing appointment scheduling and payments; (ii) electronic prescription of medication; (iii) electronic health records to store clinical information about patients, such as symptoms and treatments; and (iv) population health management through graphics that show the evolution of health parameters for a set of patients. The first two features are out of the scope of the current paper, so we will not address them further.

The software applications that provide features (iii) and (iv) and that are the most used in Portugal, are: Glintt HS[5], iMed[6] and MedicineOne[7]. Two additional software applications used in Brazil (therefore, also supporting Portuguese) are IClinic[8] and HiDoctor[9]. Other software applications widely used according to the Medical Economics web site[10] are: CareAware[11], and eClinicalWorks[12], among others. We analyzed the functionalities provided by these tools based on the

---

[5] http://www.glintt.com/.

[6] https://www.imed.com.pt/.

[7] http://www.medicineone.net/.

[8] https://iclinic.com.br/.

[9] https://hidoctor.com.br/.

[10] http://medicaleconomics.modernmedicine.com/medical-economics/content/tags/top100ehrs/top-100-ehr-companies-part-1-4.

[11] https://store.cerner.com/items/2.

[12] https://www.eclinicalworks.com/.

information available in the corresponding web sites and on the experience of our team's physician in his daily activities in Portuguese hospitals.

The result of our analysis of these applications is as follows. First, the forms and screens of these applications' user interfaces show all possible medical information (e.g., examinations or symptoms) independently of the medical specialty. This kind of user interface adds an overhead to the daily activity of physicians, because they waste time searching for the most adequate information field to fill in. Second, most applications typically support many free-form fields to be filled in by physicians. Consequently, unstructured (or textual) data is stored; this prevents the use of data mining algorithms on the data to discover knowledge for assisting physicians in making decisions. Third, none of the analyzed tools supports internationally-accepted questionnaires for assessing patient's health status (for example, IPSS in Urology) in electronic format. The electronic support for these questionnaires is important because it allows patients to answer them comfortably at home (or wherever they wish). As a consequence, the computation of a score that assesses their health condition may occur sooner. Very few of these tools (for example, careAware) support the visualization of a patient's health status evolution over time. Finally, only a few of these tools (e.g., eClinicalWorks) offers a functionality for analyzing the evolution of the health status of a set of patients (i.e., a population) over time.

In terms of research literature, we present two types of work: (i) those whose goal is to apply data mining algorithms to large medical datasets to extract useful information, and (ii) those whose goal is to apply visualization techniques to facilitate the exploration of data and analysis results by the end users. In the first group, we highlight the work by Donzé et al. [3] that provides an example of how Logistic Regression can be used to make predictions about a patient outcome. Che et al. [1] show how Bayes networks can be used to detect temporal patterns in time-series data, which is useful to predict the evolution of a patient over time. Finally, Najjar et al.'s paper [6] describes a powerful clustering approach that can be used to find groups of similar patients.

In the second group of research projects, LifeLines [8] was a seminal work developed for visualization of a patient's medical history. LifeLines2 [10,12,13] is a system designed to search and explore event sequences in temporal categorical data from multiple records of patient data. Gravi++ [4] is an application that clusters patients according to the similarity of their ordinal categorical attribute values. Furthermore, Gravi++ supports the visualization of the evolution of patients through time with an animation controllable by the user.

To conclude, none of the available commercial systems fulfills the needs of physicians in terms of medical data entry and analysis for clinical support. Existing research works provide important contributions in terms of the application of data mining algorithms or visualization techniques that can be integrated into our solution.

6   Conclusions

We described Umedicine, a software system that supports clinical activity by keeping medical information up to date and stored in an organized way. It provides easy, quick and always-on access to a large amount of clinical data. The medical information is kept persistently in an electronic and structured format. It provides secure authentication for four types of user and a specific user interface for each type of user. Additionally, the system includes a clustering algorithm suited to discover groups of patients with similar medical histories and a visualization mechanism to explore the clustering results, which shows the main characteristics, treatments and statistics of the patient clusters.

Experimental evaluation revealed that users were able to understand the discovered groups of patients and the reasons (characteristics) why they were in the same group. Users were also able to identify the situations where patients improved or worsened. In spite of these results, our solution still needs some improvements. One is to redesign the options of the visualization mechanism for an easier understanding by users. The data analysis capabilities also have plenty of room for expansion: other machine learning algorithms that take into account the different widths of medical time series can be integrated into the system to suggest treatments or predict treatment outcomes based on the patient's history. To deploy Umedicine in a real hospital environment, two additional issues should be taken into account: (i) the need for Application Programming Interfaces to enable the integration with third-party systems and (ii) the compliance with international standards for representing health data, like SNOMED[13] or LOINC[14]. Finally, it is important to test the application in a real clinical environment, with actual physicians as users and data from real patients. This experience will certainly guide future decisions regarding new functionalities to be integrated in the system and eventually determine the development lines that will be followed.

Acknowledgements. This work was supported by national funds through Fundação para a Ciência e a Tecnologia (FCT) with reference UID/CEC/50021/2013 (INESC-ID) and UID/CEC/00408/2013 (LASIGE).

References

1. Che, Z., Kale, D., Li, W., Bahadori, M.T., Liu, Y.: Deep Computational Phenotyping. In: Proceedings of the 21st ACM SIGKDD International Conference on Knowledge Discovery and Data Mining, KDD 2015, pp. 507–516. ACM, New York (2015)
2. Christensen, T., Grimsmo, A.: Instant availability of patient records, but diminished availability of patient information: a multi-method study of GP's use of electronic patient records. BMC Med. Inform. Decis. Mak. **8**, 12 (2008)

---

[13] http://www.snomed.org.
[14] http://loinc.org.

3. Donzé, J., Aujesky, D., Williams, D., Schnipper, J.L.: Potentially avoidable 30-day hospital readmissions in medical patients: derivation and validation of a prediction model. JAMA Internal Med. **173**(8), 632 (2013)

4. Hinum, K., Miksch, S., Aigner, W., Ohmann, S., Popow, C., Pohl, M., Rester, M.: Gravi++: interactive information visualization to explore highly structured temporal data. J. UCS **11**(11), 1792–1805 (2005)

5. Madeira, S.C., Teixeira, M.C., Sa-Correia, I., Oliveira, A.L.: Identification of regulatory modules in time series gene expression data using a linear time biclustering algorithm. IEEE/ACM Trans. Comput. Biol. Bioinform. **7**(1), 153–165 (2010)

6. Najjar, A., Gagné, C., Reinharz, D.: Two-step heterogeneous finite mixture model clustering for mining healthcare databases. In: 2015 IEEE International Conference on Data Mining (ICDM), pp. 931–936, November 2015

7. Peeters, R.: The maximum edge biclique problem is NP-complete. Discrete Appl. Math. **131**(3), 651–654 (2003)

8. Plaisant, C., Mushlin, R., Snyder, A., Li, J., Heller, D., Shneiderman, B.: Life-Lines: using visualization to enhance navigation and analysis of patient records. In: Proceedings of the AMIA Symposium, pp. 76–80 (1998)

9. Prelić, A., Bleuler, S., Zimmermann, P., Wille, A., Bühlmann, P., Gruissem, W., Hennig, L., Thiele, L., Zitzler, E.: A systematic comparison and evaluation of biclustering methods for gene expression data. Bioinformatics **22**(9), 1122–1129 (2006)

10. Rind, A.: Interactive information visualization to explore and query electronic health records. Found. Trends® Hum.-Comput. Interact. **5**(3), 207–298 (2013)

11. Smith, R.: What clinical information do doctors need? BMJ **313**(7064), 1062–1068 (1996)

12. Wang, T.D., Plaisant, C., Quinn, A.J., Stanchak, R., Murphy, S., Shneiderman, B.: Aligning temporal data by sentinel events: discovering patterns in electronic health records. In: Proceedings of the SIGCHI Conference on Human Factors in Computing Systems, CHI 2008, pp. 457–466. ACM, New York (2008)

13. Wang, T.D., Plaisant, C., Shneiderman, B., Spring, N., Roseman, D., Marchand, G., Mukherjee, V., Smith, M.: Temporal summaries: supporting temporal categorical searching, aggregation and comparison. IEEE Trans. Visual Comput. Graph. **15**(6), 1049–1056 (2009)

Association Rule Learning and Frequent Sequence Mining of Cancer Diagnoses in New York State

Yu Wang[1] and Fusheng Wang[1,2(✉)]

[1] Department of Computer Science, Stony Brook University,
Stony Brook, NY 11794, USA
yuwang4@cs.stonybrook.edu, fusheng.wang@stonybrook.edu
[2] Department of Biomedical Informatics, Stony Brook University,
Stony Brook, NY 11794, USA

Abstract. Analyzing large scale diagnosis histories of patients could help to discover comorbidity or disease progression patterns. Recently, open data initiatives make it possible to access statewide patient data at individual level, such as New York State SPARCS data. The goal of this study is to explore frequent disease co-occurrence and sequence patterns of cancer patients in New York State using SPARCS data. Our collection includes 18,208,830 discharge records from 1,565,237 patients with cancer-related diagnoses during 2011–2015. We use Apriori algorithm to discover top disease co-occurrences for common cancer categories based on support. We generate top frequent sequences of diagnoses with at least one cancer related diagnosis from patients' diagnosis histories using the cSPADE algorithm. Our data driven approach provides essential knowledge to support the investigation of disease co-occurrence and progression patterns for improving the management of multiple diseases.

Keywords: Association rule learning · Sequence mining · SPARCS

1 Introduction

Disease co-occurrence, which means that two or more diseases co-occur within one patient [1], is a popular topic in public health studies. It sometimes represents comorbidity or multimorbidity and can suggest interactions between different risk factors like diagnoses, treatments and procedures [1,2]. Data mining and machine learning techniques are widely applied to public health domain to discover disease co-occurrences. For example, statistical methods can be used to measure the association between two different diagnoses [3], and structure learning models like Bayesian Network are used to analyze interactions in disease co-occurrence patterns [2]. Disease co-occurrences can also be identified by computing diseases that co-occur most frequently using Apriori-like algorithms [4]. Patterns and features discovered from comorbidities could provide a foundation

E. Begoli et al. (Eds.): DMAH 2017, LNCS 10494, pp. 121–135, 2017.
DOI: 10.1007/978-3-319-67186-4_10

for creating predictive models [5]. For example, comorbidities are informative features in predicting readmission risk of certain diseases [1].

Although disease co-occurrence is essential in studying correlations among different diseases, it fails to suggest temporal trends of diagnoses as information on the order in which diseases occur is not available. It therefore cannot reveal disease progression. Sequential data mining, which considers the order of data elements, has been used to detect temporal trends of various diseases. For example, windowing, episode rules and inductive logic programming are used to extract frequent sequential patterns of cardiovascular diseases [6]. Aggregate values and time intervals from health records are used as features to cluster patients into different cohorts [7]. Wavelet functions can help to analyze time series in healthcare data of patients with diabetes [8]. However, most of these methods are value-based, they use values from laboratory tests or other healthcare records to generate results. In our study, we adopt a sequence mining method that uses diagnosis codes (class labels) to study disease progression from patients' diagnosis histories.

Recently, open data initiatives from governments collect and make available large amounts of healthcare data, and provide a unique opportunity to study disease comorbidities and sequential patterns. They are attractive to researchers working on public health studies because of their completeness and inexpensive nature [9,10]. Such data are extensively used in healthcare research, such as prevention and detection of diseases, studying comorbidity and mortality, and advancing interventions, therapies and treatments [9]. They can also be combined with multiple data sources to serve different purposes, such as studying disease patterns and improving healthcare quality among different cohorts. For instance, predicting asthma-related emergency department visits [11] and analyzing temporal patterns of in-hospital falls among elderly patients [12].

As part of New York State's open data initiative, New York State Statewide Planning and Research Cooperative System (SPARCS) collects patient-level information on discharge records from hospitals, which contains patients' diagnosis, procedure and demographic information for over 35 years [13]. SPARCS is now widely applied to public health studies in New York State [14,15], such as correlations between various factors and outcomes of patients who suffer from different diseases [16–19], associations of different patient characteristics, diseases and treatments [16,20]. SPARCS is also used to discover temporal or spatial patterns of emergency department visits before, during and after Hurricane Sandy [21,22]. Researchers can benefit from SPARCS data by leveraging the long patient-level diagnosis histories, such as conducting population-based studies [23] and assessing completeness of disease reporting [24]. Patient-level longitudinal data can also embrace other data sources like drug exposure profiles and genetics data to study patterns in different cohorts [5].

The objective of this study is to find association rules (i.e., co-occurrences of diseases) and frequent sequence patterns from diagnosis histories of cancer patients in New York State using SPARCS data. Association rules learning of multiple diseases could imply comorbidities, while sequence patterns of diseases could indicate disease progression. We extract all discharge records of patients with at least one cancer-related diagnosis code, and convert the ninth and tenth revision of International Classification of Diseases (ICD-9 and ICD-10) diagnosis codes to single-level Clinical Classifications Software (CCS) diagnosis categories. The CCS cancer categories are used as disease labels in our work. We use Apriori algorithm for association rules learning to find potential comorbidities using multiple diagnoses from individual visits and cSPADE algorithm for frequent sequence mining to identify frequent disease sequence patterns from full discharge histories of patients in each cohort. We perform the studies by using only primary diagnoses and using all diagnoses (including secondary ones), to generate different patterns. We present the results based on several common cancer types, and we believe that the results will provide essential data and knowledge for clinical researchers to further investigate comorbidities and disease progression for improving the management of multiple diseases.

2   Methods

Using data mining and machine learning methods to study patients' profiles can help researchers to study comorbidities and disease progression [5]. Our objective is to conduct a patient-level longitudinal study using SPARCS data to discover frequent disease co-occurrence and sequence patterns. We first convert ICD-9

Table 1. Cancer-related CCS diagnosis categories and descriptions.

CCS	Description	CCS	Description
11	Cancer of head and neck	29	Cancer of prostate
12	Cancer of esophagus	30	Cancer of testis
13	Cancer of stomach	31	Cancer of other male genital organs
14	Cancer of colon	32	Cancer of bladder
15	Cancer of rectum and anus	33	Cancer of kidney and renal pelvis
16	Cancer of liver and intrahepatic bile duct	34	Cancer of other urinary organs
17	Cancer of pancreas	35	Cancer of brain and nervous system
18	Cancer of other GI organs; peritoneum	36	Cancer of thyroid
19	Cancer of bronchus; lung	37	Hodgkin's disease
20	Cancer; other respiratory and intrathoracic	38	Non-Hodgkin's lymphoma
21	Cancer of bone and connective tissue	39	Leukemias
22	Melanomas of skin	40	Multiple myeloma
23	Other non-epithelial cancer of skin	41	Cancer; other and unspecified primary
24	Cancer of breast	42	Secondary malignancies
25	Cancer of uterus	43	Malignant neoplasm without specification of site
26	Cancer of cervix	44	Neoplasms of unspecified nature or uncertain behavior
27	Cancer of ovary	45	Maintenance chemotherapy; radiotherapy
28	Cancer of other female genital organs		

and ICD-10 diagnosis codes to CCS diagnosis categories, and then use Apriori and cSPADE algorithms to identify patterns using these high-level categories. We only focus on histories of patients who have at least one of the cancer-related CCS diagnosis categories (Table 1).

2.1 Data Sources

We use SPARCS data and obtain histories of 21,466,868 patients from 97,849,071 discharge records during 2011–2015. Discharge records with all four kinds of claim types (i.e. inpatient, outpatient, ambulatory surgery and emergency department) are used to get a full history of each patient. Table 2 shows patient characteristics of our experiment data.

Table 2. Statistics of patient characteristics for selected cancer types.

Patient characteristics		Cancer						
		Lung and bronchus	Rectum and anus	Pancreas	Liver[a]	Non-Hodgkin's lymphoma	Prostate	Breast
Total number of patients		121,108	40,865	25,424	28,244	75,824	198,067	300,929
Age	<65	43,002	21,696	9,858	14,990	38,366	54,760	152,393
	65–74	38,120	9,492	7,333	7,304	17,513	64,154	69,886
	75–85	30,336	6,865	5,821	4,636	14,105	55,660	51,501
	>85	9,650	2,812	2,412	1,314	5,840	23,493	27,149
Sex	Male	58,320	20,818	12,694	17,348	39,102	198,067	3,919
	Female	62,785	20,043	12,729	10,896	36,719	0	297,004
	Unknown	3	4	1	0	3	0	6
Race	White	88,718	27,342	17,107	16,224	54,247	133,541	207,660
	Black or African American	12,490	5,022	3,369	3,979	6,948	31,120	34,221
	Native American or Alaskan Native	237	110	40	77	153	416	606
	Asian	3,713	1,428	800	2,062	1,647	3,000	8,483
	Native Hawaiian or Other Pacific Islander	224	64	35	50	101	375	532
	Other Race	13,934	6,212	3,710	5,446	11,358	26,512	43,372
	Unknown	1,792	687	363	406	1,370	3,103	6,055
Ethnicity	Spanish/Hispanic Origin	7,160	3,541	1,960	3,270	6,014	14,285	22,183
	Not of Spanish/Hispanic Origin	108,808	35,437	22,467	23,879	66,534	175,028	264,771
	Unknown	5,140	1,887	997	1,095	3,276	8,754	13,975

[a] Liver includes intrahepatic bile duct.

There are 25 data elements used to record ICD diagnosis codes of each hospital visit in SPARCS. The first diagnosis code is the primary diagnosis code that represents a main reason for a patient's hospital visit, the rest are secondary diagnosis codes that represent conditions coexist during that hospital visit. All ICD-9 and ICD-10 diagnosis codes are converted to their corresponding single-level CCS diagnosis categories, i.e. primary diagnosis categories and secondary

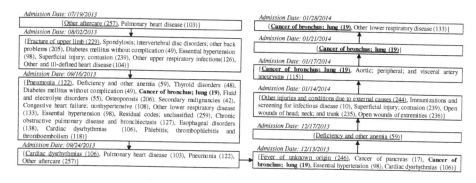

Fig. 1. Diagnoses sequence of a patient with lung and bronchus cancer.

diagnosis categories. These high-level diagnosis categories are used to represent disease diagnoses to reduce dimensionality in data mining. We study patients with cancer diagnosis categories only. For each cancer category, patients whoever have at least one discharge record containing the cancer-related diagnosis information are selected into the cohort. There are 1,565,237 cancer patients and 18,208,830 history discharge records used in this study. Each patient's discharge records are grouped together using an encrypted unique patient identifier in SPARCS. Due to the length limit of this paper, we select seven types of cancers with high incident rates, which are consistent with the statistics by American Cancer Society [25], to present our results.

For each patient, discharge records are ordered by admission dates such that all CCS diagnosis categories on the same admission date form an element, and all elements are ordered to constitute a sequence (Fig. 1). Discharge records contain AIDS/HIV or abortion diagnoses are deleted from our experiment data because the admission dates are redacted and we cannot decide their positions in a sequence. An example of diagnoses sequence of a patient in cohort with lung and bronchus cancer is shown in Fig. 1. CCS diagnosis category descriptions reported on the same admission date are listed in brackets and form an element. The corresponding CCS category labels are marked in the parentheses following the descriptions. Admission dates are marked on top of each corresponding element. The primary diagnosis category of each element is underlined. CCS category that represents the targeted cancer (i.e., lung and bronchus cancer) is highlighted in bold.

2.2 Apriori Algorithm: Identifying Disease Co-occurrence Patterns

Association rule learning is a rule-based machine learning approach and is usually used to identify co-occurrences or temporal patterns between diseases in clinical domain [4]. In this study, we adopt Apriori algorithm [26] to identify disease co-occurrence patterns among each cohort. Only elements with targeted cancer CCS diagnosis categories are selected, and both primary and secondary diagnosis categories are used in our experiment. For instance, for the sequence illustrated in Fig. 1, elements where the targeted cancer CCS diagnosis categories are highlighted in bold are used.

Apriori algorithm discovers frequent disease co-occurrences by comparing their supports with a user-specified minimum support threshold. In Fig. 1, for example, if the support of pattern "{Cancer of bronchus; lung (19), Other lower respiratory disease (133)}" is 15%, it means that 15% of the elements in this cohort have this disease co-occurrence pattern. If the minimum support threshold is greater than 15%, this pattern will not be identified. However, if the minimum support threshold is set smaller than 15%, the pattern will be detected.

2.3 cSPADE Algorithm: Discovering Frequent Sequence Patterns

Because ICD diagnosis codes are the only data elements available in SPARCS that contain patient-level disease information, we can use frequent sequence mining [27] technique to find frequent disease sequence patterns among different cohorts. Since diagnosis codes are strictly ordered in sequences, the results might reveal disease progression. We use cSPADE algorithm [27] to discover frequent disease sequence patterns in different cohorts. We experiment on complete patient sequences with two settings: one is using only primary diagnosis categories, the other one is using both primary and secondary diagnosis categories. Figure 1 is an example of a complete patient sequence consists of both primary and secondary diagnosis categories. The length of a sequence pattern is the total number of elements in this sequence. There are 10 elements in the sequence in Fig. 1, thus it is a length-10 sequence.

cSPADE algorithm also works by comparing the support of a sequence pattern with the minimum support threshold. Multiple occurrences of a pattern in the same sequence is counted only once. For example, length-2 sequence pattern "{Cancer of bronchus; lung (19), Other lower respiratory disease (133)} → {Cardiac dysrhythmias (106)}" appears twice in Fig. 1, but this pattern will be counted only once in this sequence when calculating the support of this pattern. If the support of this sequence pattern is 15%, it means that the fraction of sequences containing this pattern in the targeted cohort is 15%. If the minimum support threshold is smaller than 15%, this sequence pattern is selected; otherwise the pattern is pruned in the searching results.

3 Results

We present the top five frequent disease co-occurrence and sequence patterns ranked by their supports in each cohort. Some meaningless results, such as patterns containing identical diagnosis categories, CCS diagnosis categories that represent unspecific disease groups or serve administrative purposes, patterns with length one and patterns irrelevant to targeted cancers, are filtered out when refining experiment results. We choose to present length-2 disease sequences in our experiment results, because longer disease sequence patterns obtained in our experiments usually contain repeated diagnosis categories that represent follow-up visits rather than disease progression.

Frequent disease co-occurrence patterns are presented in Table 3, and the results are generated using both primary and secondary diagnosis categories. Table 4 presents frequent disease sequence patterns discovered using only primary diagnosis categories. Table 5 demonstrates frequent disease sequence patterns identified using both primary and secondary diagnosis categories.

4 Discussion

4.1 Common CCS Categories in Different Cohorts

We can learn from Tables 3 and 5 that essential hypertension is the most frequent CCS diagnosis category among all results of either frequent disease co-occurrence or sequence patterns. However, essential hypertension appears in only three sequences in Table 4. This might because of the difference between primary diagnosis codes and secondary diagnosis codes in SPARCS data. Results in Tables 3 and 5 are generated using both primary and secondary diagnosis categories, but patterns in Table 4 are discovered using primary diagnosis categories only. Since primary diagnosis codes usually represent one major reason for a hospital visit and secondary diagnosis codes imply conditions that coexist during this visit, a combination of primary and secondary diagnosis codes usually contain richer diagnosis information. Perhaps cancers are more likely to be diagnosed with in the elderly and essential hypertension tend to be popular among old people, thus patients with cancer diagnoses could usually have essential hypertension. Combining primary and secondary diagnosis codes can help us easily detect this pattern. Disorders of lipid metabolism is another diagnosis category that is frequent in both Tables 3 and 5, while unseen in Table 4. The underlying theory might be similar.

4.2 Disparities Between Primary and Secondary Diagnosis Codes

Tables 4 and 5 both present frequent disease sequence patterns among different cohorts, while Table 4 shows the results produced using primary diagnosis categories only and Table 5 demonstrates results using both primary and secondary diagnosis categories. Frequent disease sequence patterns among same cohorts in these two tables are quite different. Disparities between Tables 4 and 5 could imply that either primary diagnosis codes or secondary diagnosis codes may be or may not be useful in finding potentially meaningful disease sequence patterns. Since primary diagnosis codes usually represent the main reason of a hospital visit, these codes are supposed to be good indicators of a patient's condition at admission. However, secondary diagnosis codes simply represent conditions that coexist in the same hospital visit, they might not be able to accurately represent a patient's condition responsible for that hospital visit. Thus, secondary diagnosis codes could be less meaningful information in this study. This can be justified by comparing results in Tables 4 and 5.

Table 3. Frequent disease co-occurrences for selected cancers, using both primary and secondary diagnosis categories.

Lung and bronchus cancer	
Top five frequent disease co-occurrences	Support
1 Essential hypertension	0.2496
2 Screening and history of mental health and substance abuse codes	0.2271
3 Chronic obstructive pulmonary disease and bronchiectasis	0.1948
4 Disorders of lipid metabolism	0.1595
5 Coronary atherosclerosis and other heart disease	0.1136

Rectum and anus cancer	
Top five frequent disease co-occurrences	Support
1 Essential hypertension	0.1908
2 Disorders of lipid metabolism	0.1088
3 Screening and history of mental health and substance abuse codes	0.0994
4 Deficiency and other anemia	0.0896
5 Diabetes mellitus without complication	0.0796

Pancreas cancer	
Top five frequent disease co-occurrences	Support
1 Essential hypertension	0.2216
2 Diabetes mellitus without complication	0.1456
3 Fluid and electrolyte disorders	0.1260
4 Disorders of lipid metabolism	0.1234
5 Deficiency and other anemia	0.1065

Liver and intrahepatic bile duct cancer	
Top five frequent disease co-occurrences	Support
1 Essential hypertension	0.2435
2 Hepatitis	0.2275
3 Diabetes mellitus without complication	0.1494
4 Screening and history of mental health and substance abuse codes	0.1340
5 Fluid and electrolyte disorders	0.1247

Non-Hodgkin's lymphoma	
Top five frequent disease co-occurrences	Support
1 Essential hypertension	0.1874
2 Deficiency and other anemia	0.1224
3 Disorders of lipid metabolism	0.1213
4 Screening and history of mental health and substance abuse codes	0.0910
5 Diabetes mellitus without complication	0.0835

Prostate cancer	
Top five frequent disease co-occurrences	Support
1 Essential hypertension	0.3292
2 Disorders of lipid metabolism	0.2384
3 Coronary atherosclerosis and other heart disease	0.1739
4 Disorders of lipid metabolism, Essential hypertension	0.1521
5 Screening and history of mental health and substance abuse codes	0.1392

Breast cancer	
Top five frequent disease co-occurrences	Support
1 Essential hypertension	0.2159
2 Disorders of lipid metabolism	0.1302
3 Disorders of lipid metabolism, Essential hypertension	0.0851
4 Diabetes mellitus without complication	0.0829
5 Screening and history of mental health and substance abuse codes	0.0792

Table 4. Frequent sequence patterns for selected cancers, using primary diagnosis categories only.

Lung and bronchus cancer	
Top five frequent sequence patterns	Support
1 {Chronic obstructive pulmonary disease and bronchiectasis}→{Lung and bronchus cancer}	0.0700
2 {Pneumonia}→{Lung and bronchus cancer}	0.0578
3 {Lung and bronchus cancer}→{Pneumonia}	0.0567
4 {Lung and bronchus cancer}→{Chronic obstructive pulmonary disease and bronchiectasis}	0.0524
5 {Lung and bronchus cancer}→{Septicemia}	0.0520

Rectum and anus cancer	
Top five frequent sequence patterns	Support
1 {Rectum and anus cancer}→{Colon cancer}	0.1323
2 {Colon cancer}→{Rectum and anus cancer}	0.1206
3 {Rectum and anus cancer}→{Complications of surgical procedures or medical care}	0.0521
4 {Gastrointestinal hemorrhage}→{Rectum and anus cancer}	0.0509
5 {Abdominal pain}→{Rectum and anus cancer}	0.0479

Pancreas cancer	
Top five frequent sequence patterns	Support
1 {Pancreatic disorders}→{Pancreas cancer}	0.1256
2 {Abdominal pain}→{Pancreas cancer}	0.0994
3 {Biliary tract disease}→{Pancreas cancer}	0.0914
4 {Pancreas cancer}→{Septicemia}	0.0794
5 {Pancreas cancer}→{Abdominal pain}	0.0618

Liver and intrahepatic bile duct cancer	
Top five frequent sequence patterns	Support
1 {Hepatitis}→{Liver and intrahepatic bile duct cancer}	0.0711
2 {Abdominal pain}→{Liver and intrahepatic bile duct cancer}	0.0688
3 {Liver and intrahepatic bile duct cancer}→{Hepatitis}	0.0586
4 {Liver and intrahepatic bile duct cancer}→{Septicemia (except in labor)}	0.0528
5 {Biliary tract disease}→{Liver and intrahepatic bile duct cancer}	0.0458

Non-Hodgkin's lymphoma	
Top five frequent sequence patterns	Support
1 {Lymphadenitis}→{Non-Hodgkin's lymphoma}	0.0458
2 {Non-Hodgkin's lymphoma}→{Septicemia (except in labor)}	0.0458
3 {Non-Hodgkin's lymphoma}→{Deficiency and other anemia}	0.0433
4 {Deficiency and other anemia}→{Non-Hodgkin's lymphoma}	0.0415
5 {Non-Hodgkin's lymphoma}→{Diseases of white blood cells}	0.0368

Prostate cancer	
Top five frequent sequence patterns	Support
1 {Prostate cancer}→{Genitourinary symptoms and ill-defined conditions}	0.0424
2 {Genitourinary symptoms and ill-defined conditions}→{Prostate cancer}	0.0401
3 {Essential hypertension}→{Prostate cancer}	0.0323
4 {Prostate cancer}→{Essential hypertension}	0.0292
5 {Spondylosis; intervertebral disc disorders; other back problems}→{Prostate cancer}	0.0283

Breast cancer	
Top five frequent sequence patterns	Support
1 {Nonmalignant breast conditions}→{Breast cancer}	0.0965
2 {Breast cancer}→{Nonmalignant breast conditions} .	0.0796
3 {Spondylosis; intervertebral disc disorders; other back problems}→{Breast cancer}	0.0353
4 {Breast cancer}→{Spondylosis; intervertebral disc disorders; other back problems}	0.0331
5 {Essential hypertension}→{Breast cancer}	0.0295

Table 5. Frequent sequence patterns for selected cancers, using both primary and secondary diagnosis categories.

Lung and bronchus cancer	
Top five frequent sequence patterns	Support
1 {Essential hypertension}→{Lung and bronchus cancer}	0.5377
2 {Screening and history of mental health and substance abuse codes}→{Lung and bronchus cancer}	0.5240
3 {Lung and bronchus cancer}→{Essential hypertension}	0.4508
4 {Lung and bronchus cancer}→{Screening and history of mental health and substance abuse codes}	0.4350
5 {Disorders of lipid metabolism}→{Lung and bronchus cancer}	0.4044

Rectum and anus cancer	
Top five frequent sequence patterns	Support
1 {Essential hypertension}→{Rectum and anus cancer}	0.4338
2 {Rectum and anus cancer}→{Colon cancer}	0.4273
3 {Rectum and anus cancer}→{Essential hypertension}	0.4218
4 {Colon cancer}→{Rectum and anus cancer}	0.3863
5 {Disorders of lipid metabolism}→{Rectum and anus cancer}	0.2931

Pancreas cancer	
Top five frequent sequence patterns	Support
1 {Essential hypertension}→{Pancreas cancer}	0.5440
2 {Pancreas cancer}→{Essential hypertension}	0.4156
3 {Disorders of lipid metabolism}→{Pancreas cancer}	0.3857
4 {Essential hypertension}→{Essential hypertension, Pancreas cancer}	0.3767
5 {Fluid and electrolyte disorders}→{Pancreas cancer}	0.3718

Liver and intrahepatic bile duct cancer	
Top five frequent sequence patterns	Support
1 {Essential hypertension}→{Liver and intrahepatic bile duct cancer}	0.5116
2 {Liver and intrahepatic bile duct cancer}→{Essential hypertension}	0.4205
3 {Liver and intrahepatic bile duct cancer}→{Fluid and electrolyte disorders}	0.3510
4 {Screening and history of mental health and substance abuse codes}→{Liver and intrahepatic bile duct cancer}	0.3439
5 {Fluid and electrolyte disorders}→{Liver and intrahepatic bile duct cancer}	0.3271

Non-Hodgkin's lymphoma	
Top five frequent sequence patterns	Support
1 {Essential hypertension}→{Non-Hodgkin's lymphoma}	0.4237
2 {Non-Hodgkin's lymphoma}→{Essential hypertension}	0.3911
3 {Disorders of lipid metabolism}→{Non-Hodgkin's lymphoma}	0.3128
4 {Deficiency and other anemia}→{Non-Hodgkin's lymphoma}	0.2925
5 {Non-Hodgkin's lymphoma}→{Deficiency and other anemia}	0.2920

Prostate cancer	
Top five frequent sequence patterns	Support
1 {Essential hypertension}→{Prostate cancer}	0.4849
2 {Prostate cancer}→{Essential hypertension}	0.4667
3 {Disorders of lipid metabolism}→{Prostate cancer}	0.3707
4 {Prostate cancer}→{Disorders of lipid metabolism}	0.3654
5 {Genitourinary symptoms and ill-defined conditions}→{Prostate cancer}	0.2156

Breast cancer	
Top five frequent sequence patterns	Support
1 {Essential hypertension}→{Breast cancer}	0.3958
2 {Breast cancer}→{Essential hypertension}	0.3887
3 {Disorders of lipid metabolism}→{Breast cancer}	0.2804
4 {Breast cancer}→{Disorders of lipid metabolism}	0.2779
5 {Nonmalignant breast conditions}→{Breast cancer}	0.2177

For patients with lung and bronchus cancer in Table 4, the most frequent sequences mainly consist of respiratory system diseases, such as pneumonia and chronic obstructive pulmonary disease and bronchiectasis. But there is no respiratory system disease in the top five frequent disease sequence patterns among the same patient cohort in Table 5. Another typical cohort is patients with liver and intrahepatic bile duct cancer. We can learn from Table 4 that patients in this cohort sometimes expose themselves to hepatitis or biliary tract disease. However, such patterns are not available in Table 5. Also for patients with Non-Hodgkin's lymphoma, frequent sequence patterns shown in Tables 4 and 5 are quite different. Only results in Table 4 capture the existence of lymphadenitis and disease of white blood cells. Also, the most frequent disease sequence patterns among this cohort in Table 4 all consist of immune system diseases.

4.3 Frequent Disease Co-occurrence Patterns Versus Frequent Disease Sequence Patterns

One major difference between disease sequence and co-occurrence patterns is that the orders of diagnoses are taken into consideration in a disease sequence, while disease co-occurrences simply represent different diagnoses that occur simultaneously. Disease sequence pattern can therefore be a potential indicator of disease progression. Since the order of two different diagnosis categories is the major factor to consider when tracking disease progression, we retain a frequent disease sequence pattern in the results, if its elements are reversed in another top frequent disease sequence pattern.

For instance, sequence patterns "{Rectum and anus cancer} → {Colon cancer}" and "{Colon cancer} → {Rectum and anus cancer}" are both kept in Table 4. The former has support 0.1323, which is slightly greater than the latter (0.1206). Perhaps it is because that rectum and anus cancer are more likely to develop into colon cancer, but fewer patients suffer from colon cancer can eventually have rectum and anus cancer. There could be causal relationships between the two diseases, or perhaps it is simply a result of the different mechanisms of these two types of cancers.

Another typical pattern is in disease sequences containing essential hypertension. In Table 5, for example, sequence pattern "{Essential hypertension} → {Pancreas cancer}" has support 0.5440, which is higher than the reversed sequence "{Pancreas cancer} → {Essential hypertension}" with support 0.4156. It is evident that all the sequences where essential hypertension is at the first position have higher supports than their reversed sequences. It is an interesting phenomenon that perhaps imply the progression of pancreas cancer. However, we cannot obtain any information on disease progression from disease co-occurrence patterns. For example, Table 3 shows that pattern "{Pancreas cancer, Essential hypertension}" is with the highest support among patients with pancreas cancer. It simply suggests that these two diagnoses co-occur frequently, but no information on the order in which they occur is available.

4.4 Validation of Results

Many public health studies use data from only one or a few hospitals collected in a short period of time [3, 4, 10]. However, SPARCS has been collecting more representative and comprehensive data for over 35 years, as all Article 28 facilities (i.e. hospitals, nursing homes, and diagnostic treatment centers) certified for inpatient care and all facilities providing ambulatory surgery services in New York State are required to submit inpatient or outpatient data to SPARCS [13]. We therefore have a large-scale dataset with longer patient histories that could help generate potentially meaningful results.

For disease co-occurrences (Table 3), patients with lung and bronchus cancer usually have chronic obstructive pulmonary disease and bronchiectasis observed at the same time. Since these two diseases are both respiratory system diseases, they are reasonably correlated with each other. The same applies for patients with pancreas cancer. Patients in this cohort have a risk of suffering from diabetes, as pancreas cancer and diabetes are clinically correlated [28]. Moreover, patients with liver and intrahepatic bile duct cancer also have chance to be diagnosed with hepatitis at the same time, because these two diseases are also associated with each other [28].

As for disease sequences (Table 4), many patients have pancreatic disorders (not diabetes) or biliary tract disease before being diagnosed with pancreas cancer. This might be a typical disease progression pattern in clinical studies and could help domain experts to identify pancreas cancer in the early stages. Another representative result is about Non-Hodgkin's lymphoma, because the result sequences usually consist of immune system diseases. The top sequence patterns suggest that lymphadenitis is likely to happen before Non-Hodgkin's lymphoma and disease of white blood cells is usually diagnosed after Non-Hodgkin's lymphoma.

Although secondary diagnosis codes could be redundant information on patient conditions, they are also able to produce some potentially interesting and meaningful patterns on disease progression when combined with primary diagnosis codes. For example, prostate cancer is more likely to be diagnosed after genitourinary symptoms and ill-defined conditions are identified, and breast cancer usually happens after nonmalignant breast conditions (Table 5). Since these two patterns have comparatively higher supports than other sequence patterns in the same cohorts, they could be typical patterns in clinical studies.

5 Conclusion

We employ association rule learning (Apriori algorithm) and frequent sequence mining (cSPADE algorithm) to identify frequent disease co-occurrence and sequence patterns among cancer patients using SPARCS data. Different types of diagnosis codes are utilized in our experiments. Seven cohorts where cancers are with high incident rates are selected to present the results. Our results suggest that the methods adopted can generate potentially interesting and clinically meaningful disease co-occurrence and sequence patterns. These patterns might be able to imply comorbidities and disease progression. However, due to the limitation of information that diagnosis codes can convey in SPARCS, our results contain some redundant or less meaningful patterns irrelevant to the targeted cancers. Since SPARCS is designed to serve administrative purpose to monitor and improve qualities of hospital services and data reporting, we believe our study could not only help to improve healthcare qualities provided to serve cancer patients, but also throw light upon researches using diagnosis codes in SPARCS.

Since high-level diagnosis categories contain richer but less specific diagnoses information than diagnosis codes, we can use low-level ICD-9 and ICD-10 diagnosis codes in our future researches to see if more specific and useful patterns can be extracted. We can also experiment on a cohort with one certain disease to narrow down the scope of our study and gain a deeper insight into that specific cohort.

Acknowledgments. This work is supported in part by NSF ACI 1443054, by NSF IIS 1350885 and by NSF IIP1069147.

References

1. Stiglic, G., Brzan, P.P., Fijacko, N., Wang, F., Delibasic, B., Kalousis, A., Obradovic, Z.: Comprehensible predictive modeling using regularized logistic regression and comorbidity based features. PLoS ONE **10**(12), e0144439 (2015). doi:10.1371/journal.pone.0144439
2. Lappenschaar, M., Hommersom, A., Lagro, J., Lucas, P.J.: Understanding the co-occurrence of diseases using structure learning. In: Conference on Artificial Intelligence in Medicine in Europe, pp. 135–144 (2013). doi:10.1007/978-3-642-38326-7_21
3. Munson, M.E., Wrobel, J.S., Holmes, C.M., Hanauer, D.A.: Data mining for identifying novel associations and temporal relationships with Charcot foot. J. Diabetes Res. (2014). doi:10.1155/2014/214353
4. Kost, R., Littenberg, B., Chen, E.S.: Exploring generalized association rule mining for disease co-occurrences. In: AMIA Annual Symposium Proceedings 2012, p. 1284 (2012)
5. Jensen, P.B., Jensen, L.J., Brunak, S.: Mining electronic health records: towards better research applications and clinical care. Nat. Rev. Genet. **13**(6), 395–405 (2012). doi:10.1038/nrg3208
6. Kléma, J., Nováková, L., Karel, F., Stepankova, O., Zelezny, F.: Sequential data mining: a comparative case study in development of atherosclerosis risk factors. IEEE Trans. Syst. Man Cybern. Part C (Applications and Reviews) **38**(1), 3–15 (2008). doi:10.1109/tsmcc.2007.906055
7. Baxter, R.A., Williams, G.J., He, H.: Feature selection for temporal health records. In: Pacific-Asia Conference on Knowledge Discovery and Data Mining, pp. 198–209 (2001). doi:10.1007/3-540-45357-1_24
8. Lin, W., Orgun, M.A., Williams, G.J.: Mining temporal patterns from health care data. In: International Conference on Data Warehousing and Knowledge Discovery, pp. 222–231 (2002). doi:10.1007/3-540-46145-0_22

9. Ferver, K., Burton, B., Jesilow, P.: The use of claims data in healthcare research. Open Public Health J. **2**, 11–24 (2009). doi:10.2174/1874944500902010011
10. Tyree, P.T., Lind, B.K., Lafferty, W.E.: Challenges of using medical insurance claims data for utilization analysis. Am. J. Med. Qual. **21**(4), 269–275 (2006). doi:10.1177/1062860606288774
11. Ram, S., Zhang, W., Williams, M., Pengetnze, Y.: Predicting asthma-related emergency department visits using big data. IEEE J. Biomed. Health Inform. **19**(4), 1216–1223 (2015). doi:10.1109/jbhi.2015.2404829
12. López-Soto, P.J., Smolensky, M.H., Sackett-Lundeen, L.L., De Giorgi, A., Rodríguez-Borrego, M.A., Manfredini, R., Pelati, C., Fabbian, F.: Temporal patterns of in-hospital falls of elderly patients. Nurs. Res. **65**(6), pp. 435–445 (2016). doi:10.1097/nnr.0000000000000184
13. Statewide Planning and Research Cooperative System (SPARCS). https://www.health.ny.gov/statistics/sparcs/
14. Chen, X., Wang, F.: Integrative spatial data analytics for public health studies of new york state. In: AMIA Annual Symposium Proceedings, vol. 2016, p. 391 (2016)
15. Chen, X., Wang, Y., Schoenfeld, E., Saltz, M., Saltz, J., Wang, F.: Spatio-temporal analysis for New York State SPARCS data. In: Proceedings of 2017 AMIA Joint Summits on Translational Science (2017)
16. Bekelis, K., Missios, S., Coy, S., Rahmani, R., Singer, R.J., MacKenzie, T.A.: Surgical clipping versus endovascular intervention for the treatment of subarachnoid hemorrhage patients in New York State. PLoS ONE **10**(9), e0137946 (2015). doi:10.1371/journal.pone.0137946
17. Missios, S., Bekelis, K.: Regional disparities in hospitalization charges for patients undergoing craniotomy for tumor resection in New York State: correlation with outcomes. J. Neurooncol. **128**(2), 365–371 (2016). doi:10.1007/s11060-016-2122-0
18. Bekelis, K., Missios, S., Coy, S., MacKenzie, T.A.: Scope of practice and outcomes of cerebrovascular procedures in children. Child's Nerv. Syst. **32**(11), 2159–2164 (2016). doi:10.1007/s00381-016-3114-2
19. Bekelis, K., Missios, S., Coy, S., MacKenzie, T.A.: Comparison of outcomes of patients with inpatient or outpatient onset ischemic stroke. J. Neurointerventional Surg., pp. neurintsurg-2015 (2016). doi:10.1136/neurintsurg-2015-012145
20. Dy, C.J., Lane, J.M., Pan, T.J., Parks, M.L., Lyman, S.: Racial and socioeconomic disparities in hip fracture care. J. Bone Joint Surg. Am. **98**(10), 858–865 (2016)
21. Kim, H., Schwartz, R.M., Hirsch, J., Silverman, R., Liu, B., Taioli, E.: Effect of Hurricane Sandy on Long Island emergency departments visits. Disaster Med. Public Health Preparedness **10**(03), 344–350 (2016). doi:10.1017/dmp.2015.189
22. He, F.T., De La Cruz, N.L., Olson, D., Lim, S., Seligson, A.L., Hall, G., Jessup, J., Gwynn, C.: Temporal and spatial patterns in utilization of mental health services during and after hurricane sandy: emergency department and inpatient hospitalizations in New York City. Disaster Med. Public Health Preparedness **10**(03), 512–517 (2016). doi:10.1017/dmp.2016.89
23. Hodgins, J.L., Vitale, M., Arons, R.R., Ahmad, C.S.: Epidemiology of medial ulnar collateral ligament reconstruction: a 10-year study in New York State. Am. J. Sports Med. **44**(3), 729–734 (2016). doi:10.1177/0363546515622407
24. Arakaki, L., Ngai, S., Weiss, D.: Completeness of Neisseria meningitidis reporting in New York City, 19892010. Epidemiol. Infect. **144**(11), 2374–2381 (2016). doi:10.1017/s0950268816000406
25. Cancer facts & figures 2017. American Cancer Society (2017)

26. Agrawal, R., Srikant, R.: Fast algorithms for mining association rules. In: Proceedings of the 20th International Conference on Very Large Data Bases, VLDB, vol. 1215, pp. 487–499 (1994)
27. Zaki, M.J.: Sequence mining in categorical domains: incorporating constraints. In: Proceedings of the Ninth International Conference on Information and Knowledge Management, pp. 422–429 (2000). doi:10.1145/354756.354849
28. Mayo Clinic. http://www.mayoclinic.org

Healthsurance – Mobile App for Standardized Electronic Health Records Database

Prateek Jain[1], Sagar Bhargava[1], Naman Jain[1], Shelly Sachdeva[1(✉)],
Shivani Batra[1], and Subhash Bhalla[2]

[1] Jaypee Institute of Information Technology University, Noida, India
prateekjain1112@gmail.com,
sagarbhargava3414@gmail.com, namanjain928@gmail.com,
sachdevashelly1@gmail.com, ms.shivani.batra@gmail.com
[2] University of Aizu, Fukushima 965-8580, Japan
bhalla@u-aizu.ac.jp

Abstract. With the increasing popularity of Electronic Health Records (EHRs), there arises a need to understand its importance in terms of clinical contexts for a standard based health application. Standards for semantic interoperability propose the use of archetypes for building a health application. A usual practice followed for storing of EHRs is through graphical user interfaces. Generally, user interface is static corresponding to the underlying medical concept, often made manually and are prone to errors. However, evolution in knowledge demands for dynamically generated user interfaces to reduce time, minimize cost and enhance reliability. Current research implements mobile app for standardized Electronic Health Records Database termed as HEALTHSURANCE. The application maintains its dynamic behavior through creation of graphical user interfaces at runtime by gaining knowledge from the artefacts (known as archetypes) available from standard clinical repositories (such as Clinical Knowledge Manager). This provides easy and hassle-free user operability without any need of mobile developer. A standardized format and content helps to uplift the credibility of data and maintains a uniform and specific set of constraints used to evaluate the user's health. A generic centralized database is chosen for data storage to support evolution in clinical knowledge and to handle heterogeneity of EHRs data. Implementing mobile app based on archetype paradigm avoids reimplementation of systems, migrating databases and allows the creation of future-proof systems.

Keywords: Electronic health records databases · Standardization · Mobile application · User interface · Clinical information system

1 Introduction

In healthcare domain huge amount of patients data needs to be stored. Electronic Health Records (EHRs) [1] has the capacity for greater electronic exchange since the patient can electronically share his data with any other desired party through the use of mobile healthcare application. With the advancement in technology EHRs based on mobile devices are getting popular day by day. A study revealed that users accessing

© Springer International Publishing AG 2017
E. Begoli et al. (Eds.): DMAH 2017, LNCS 10494, pp. 136–153, 2017.
DOI: 10.1007/978-3-319-67186-4_11

EHRs system using tablets and smart phones expressed higher level of overall satisfaction [16]. Thus, mobile healthcare application provides benefits in terms of faster data analyses, reliability, safety and efficacy. It also provides an interaction between patient and doctor electronically.

1.1 Electronic Health Records Standards for Semantic Interoperability

Since digitalization of EHRs is getting popular and significant in many countries, several EHRs standard has been proposed including OpenEHR [21], ISO EN 13606 [4]. Both OpenEHR and ISO EN 13606 follows dual model architecture which constitutes two layers, i.e., the Reference Model (RM) and the Archetype Model (AM) [6]. RM defines the set of entities that form the generic building blocks of the EHRs. Clinical information is defined at this level. On the other hand, archetypes (deliverable of AM) define clinical concepts in the form of structured and constrained combinations of the entities contained in the reference model [6]. Archetypes are defined by Archetype Definition Language (ADL) [7].

1.2 Archetype Definition Language

Archetype Definition Language (ADL) is the formal language for expressing archetypes [7]. ADL makes use of various blocks for defining different aspects of an archetype such as Constraint form of ADL (cADL), Data Definition form of ADL (dADL) and First Order Predicate Logic (FOPL) as shown in Fig. 1. The formalism of

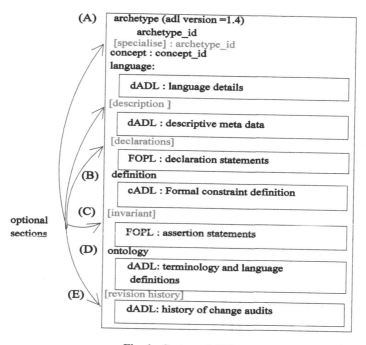

Fig. 1. Syntax of ADL

```
archetype (adl_version=1.4)
        openEHR-EHR-OBSERVATION.soap_investigations.v8
concept [at0000]--SOAP-Objective_Investigation
Language original_language = <[ISO_639-1::en]>
definition
    OBSERVATION[at0000] matches --SOAP-Objective_Investigation
    data matches {
    HISTORY[at0001] matches {    --Event Series
    events cardinality matches {1..*; unordered} matches {
    EVENT[at0002] occurrences matches {0..1} matches {--Any event
    data matches {
    ITEM_TREE[at0003] matches {--Tree
        items cardinality matches {0..*; unordered} matches {
        ELEMENT[at0004] occurrences matches {0..1} matches {   --Oximetry% value matches
        {DV_TEXT matches {*}}}
        ELEMENT[at0005] occurrences matches {0..1} matches {   -- ECG Order value matches
        {DV_BOOLEAN matches {value matches {True, False}}}}
        ELEMENT[at0006] occurrences matches {0..1} matches {   --Na Sodium result value
        matches {C_DV_QUANTITY property = <[openehr::445]> <list = <["1"] = <units =
        <"U">>>>}}}
        ELEMENT[at0007] occurrences matches {0..1} matches {   --Radiology Order value
        matches {DV_TEXT matches {
        value matches {"Chest Xray PA", "Chest Xray AP", "Pelvic Xray"}
    }}}}}}}}}}
Ontology term_definitions=<["en"]=<items=
    <["at0000"] = <text = <"SOAP-Objective_Investigation">description = <"SOAP_Investigation>
    ["at0001"] = <text = <"Event Series">description = <"@ internal @">>
    ["at0002"] = <text = <"Any event"> description = <"*">>
    ["at0003"] = <text = <"Tree">description = <"@internal @">>
    ["at0004"] = <text = <"Oximetry%">description = <"*">>
    ["at0005"] = <text = <"ECG Result">description = <"*">>
    ["at0006"] = <text = <"Na Sodium result">description = <"*">>
    ["at0007"] = <text = <"Radiology Order">description = <"*">>>>>
```

Fig. 2. Code snippet of ADL file

templates is presented by using dADL. The cADL is used to express the archetype definition, while the dADL is used to express data which appears in the language, description, ontology, and revision_history sections of an ADL archetype. The various keywords in ADL are - archetype, specialise/specialize, concept, language, description, definition, invariant, and ontology. The top-level structure of an ADL is shown in Fig. 1. An example of code snippet of ADL is shown in Fig. 2.

Converting the archetypes into user interface requires an ADL parser that can convert archetypes into a compatible language that the device is meant for (e.g., xml for android based mobile app [11] and xhtml for java based desktop application). However, the same program cannot do similar things on both devices (mobile and desktop) for factors such as processing speed, memory, and screen size. Moreover, mobile outputs display on a different activity rather than on a browser. Current research implements generation of graphical user interfaces (GUIs) from archetypes for mobile devices.

1.3   Need for Generation of Dynamic Interfaces

The semantics of data is better understood by viewing the data in the context of the user interface. Generally, software artifacts are often made manually (and are prone to errors). However, any future evolution in the medical knowledge requires changes in the existing Graphical User Interface (GUI). Thus, this research aims to generate GUIs dynamically. One GUI corresponding to one archetype is generated dynamically. Any update in medical knowledge is reflected in terms of versioning of archetypes. User can select an archetype of choice to build GUI dynamically. This reduces the burden from application builder and thus, the overall maintenance cost.

1.4   User Interfaces for Desktop and Mobile Application

The desktop is better in many ways like processor speed, screen size as presented in Table 1. Still, there are factors that make a mobile device stand apart, such as powerful sensors (that can detect location, movement, acceleration, orientation, proximity, and environmental conditions) and portability. Table 1 shows the comparison of desktop versus mobile application considering user interface generation. Statistics shows that in 2016, 62.9% of the population worldwide already owned a mobile device [15], out of which more than 49% are android users. So to increase the reachability of our mobile application, we chose android operating system as platform to build our application.

Table 1. Comparison of desktop versus mobile application.

Parameters	Desktop application	Mobile application
Screen size	Due to larger screen size, user can see most of the contents on a single page	Due to smaller screen size, user has to scroll up and down to see the contents
Font visibility	Fonts are clearly visible due to big screen size	Font visibility is not good due to small screen size
Processor speed	Interfaces can be generated fast due to high processing speed	Mobile have less processing speed due to cost and battery life so interface generation sometimes have lags
Requirement for dynamic generation	Xhtml is used to generate dynamic interfaces	Required data is extracted from xhtml to generate dynamic interfaces
Output	Output is shown in a web browser	Output is shown as a separate activity within application

Table 1 focuses on how the development of user interface in desktop and mobile differs with each other in context to parameters such as screen size, font visibility, processor speed.

As the screen size decreases, the need to zoom in contents increases. Small screen size of handheld portable devices makes data entry a difficult task. Due to smaller screens, the text size appears very small that can lead to commit mistake in filling up the data. When it comes to generating interfaces dynamically, what matters is how much time is required for its generation. The faster the processor, the less will be the time taken. As mobile devices have generally low processor than desktops, the time taken in mobile is usually higher.

The rest of the paper is organized as follows. Section 2 elucidates about the challenges faced during developing of application. Section 3 throws light on state of the art. Section 4 describes the detailed working of our application. Section 5 presents results achieved. Section 6 finally concludes current work and highlights future work.

2 Current Challenges

Considering the need of a standard based mobile application for health domain, the current research pioneers in implementing an android based application named as Healthsurance. The following challenges are encountered during creation of mobile application.

- **Dynamic Creation of GUI in Mobile** - To create android based dynamic interfaces, archetypes written in ADL are required as input. However, the constraint based templates (archetypes) are not supported in android that acts as a hindrance in generating dynamic m-forms.
- **Persistence of data** - Different GUI corresponds to different attributes of data to be persisted in database. Further, any changes in archetypes (new version) demands changes in existing database schema. Thus, a generic schema is required to support dynamic behavior of application.
- **Offline Connectivity** - The data that users fill in dynamic forms have to be stored in a standardized database located on a central web server with the help of an internet so that it is accessible to users irrespective of their location. The application must be capable to store essential data of users in offline mode and send offline stored data to central web server on regaining internet connection.
- **Enhancing usability** - For an application to be user friendly and appealing to users, it is necessary for a mobile application to have good transitions, color scheme and easy to use.

3 State-of-the-Art

Healthcare is domain of concern that can take a huge toll on lives of people. In todays' era everything is going digital and is available with us all time through mobile devices. Growing use of mobile application motivates the shift of storing health data from hospitals to local mobile devices. Usability of any application is directly affected from its user friendliness. Existing researches [2, 17–19] highlight the importance of usability for mobile applications. Current research also aims to provide high usability to a healthcare application on android based mobile devices.

Opereffa (openEHR Reference Framework and Application) [40] provides its functionality as an open source desktop based clinical application. Application proposed in current research is motivated from Opereffa and aims to improve its scope by providing functionality such as dynamic linkage to archetype repository, generation of dynamic m-forms corresponding to archetypes, and mobile based app to enhance usability. Opereffa is built on existing Java Reference Implementation, and aims to provide a workbench to have common ground for discussion to both technical and clinical communities. It provides Archetype-Template-GUI support (to avoid hard coding every time we have to develop a new form for a new disease). It has many archetypes to select which allows ad-hoc documentation. It is based on Java Server Faces (JSF) [10] for more user interface (UI) orientation [3]. For persistence, it makes use of Hibernate, a well-known java object relational mapping tool and PostgreSQL [12].

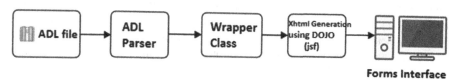

Forms Interface

Fig. 3. Dynamic generation in desktop version

Figure 3 shows the process of generation of dynamic forms in the desktop version. A file with .adl extension is given as an input to the ADL parser, which in turn gives a parsing tree and this parsing tree is converted to XHTML using wrapper class DOJO (JSF). Now with the help of XHTML, dynamic forms can easily be generated for desktop.

Authors' previous research highlights enhancing access to standardized clinical application for mobile interfaces [2]. It proposes MOpereffa for mobile devices. The use of MOpereffa for mobile devices is highlighted in [5]. In current research the authors implement an application based on interactive user interfaces for mobile device. For this purpose, our study considers Android, which is an open source technology and have a very strong java support in a way that it is platform independent. Thus, an android application does not need to be compiled every time for a new device. We aim to test our mobile application for various basic functionalities such as clerking, history, examinations, investigations, body weight, and assessment based on various archetypes.

A standard based healthcare application named AsthmaCheck [20] has been developed to store asthma related data in a generic storage. However, the application is desktop based that limits its adaptability. Moreover, AsthmaCheck stores data related to asthma only. Current application provides support for any medical concept provided an archetype is available.

Patrick et al. [24] highlighted the need of archetype based system. 10–20 basic archetypes are sufficient to build the core of the clinical information system [21–23]. To build a primary care electronic health record, approximate 100 archetypes are enough [22, 23]. Thus, authors adopt archetypes for building mobile application that persist EHRs data.

A widely adopted persistence approach is Object Relational Mapping (ORM) that creates one table in relational model corresponding to one class in object. However, adoption of ORM approach for archetype based system leads to multiple JOINS due to presence of a deep hierarchy [31]. Frade et al. [32] presents a survey that concludes that most of the existing clinical information systems (CIS) are built using relational database management system (RDBMS). It has been concluded that the impedance mismatch between the logically tree-structured archetype-based EHR and the relational model is handled in RDBMS for implementation of CIS. Wang et al. [33] proposed a new archetype relational mapping (ARM) approach that creates one relational table for each archetype. However, evolution in knowledge requires rebuilding of application. Thus, current research focuses on a generic schema. Need of a generic schema for storing standardized EHRs is highlighted in [25, 26] to support dynamic behavior of clinical application without making any modification to the database schema. Thus, current research adopts an extension of generic schema Entity Attribute Value (EAV) model known as Node+Path persistence [34] (in openEHR standard) for the purpose of data storage. EAV has been adopted for storing EHRs in many projects [27–30, 38]. However, they all provide desktop support. Current application exploits usage of EAV storage in mobile environment.

4   Architecture of Mobile App for Standardized EHRs Database

Current study aims to provide an interactive GUI for mobile devices. Developed clinical mobile application integrates various aspects of clinical recording process, such as clerking, history, examinations, investigations, and assessment. This section presents the architecture and detailed working of proposed standard based mobile clinical application.

4.1   Architecture Diagram

Figure 4 shows the architecture diagram depicting the complete functioning of our android dynamic application. It comprises of various components as listed below.

- **Web Server** - It contains all relevant archetypes in a repository and a database.
- **Archetype Repository** - It contains all the existing archetypes and new archetypes that can be uploaded.
- **Database** - A generic database to store data of EHRs when filled through mobile user interface and submitted.
- **ADL parser** - It is used to parse archetypes.
- **Dynamic generator** - It is used to create dynamic interfaces/MForms.
- **Mobile Device** - The device with android operating system used to display MForms.

The entire process can be summed as follows:-

The application requires archetypes to be saved on the local storage of the **mobile device**. These files are downloaded from the **archetype repository** situated in the central **web server**. If any new archetype is added in the archetype repository or if some or all the archetypes are missing from the device's memory, the application auto-detects and downloads it into the device's memory to carry on the process. The archetypes in the storage can be browsed and selected in order to parse via **adl parser** and the **dynamic generator** creates an interface designed specifically for the android

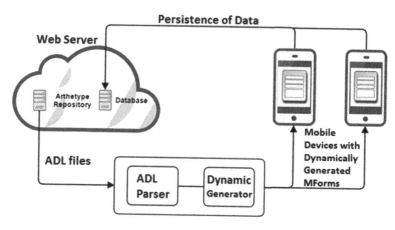

Fig. 4. Architecture diagram

devices corresponding to that archetype. The user then needs to fill the form. The data filled is saved as standardized EHRs database (in MySQL **database**) on an online web server, if there is internet connectivity. The detailed working of the proposed mobile application is described in next section.

4.2 Detailed Working of Android Based Mobile Application

Figure 5 shows comprehensive details of the working of mobile app 'Healthsurance'. It consists of three phases, i.e., handling frequent schema evolution, automatic generation of M-forms and generic persistence of EHRs database.

4.2.1 Phase 1 (Handling Frequent Schema Evolution)

Frequent schema evolution has been handled through archetypes. It has been implemented via auto-detection and auto expansion of ADL files. The onset of this phase deals with downloading archetypes from an online archetype repository such as Clinical Knowledge Manager (CKM) [14] and is stored in the local storage of device. Due to the dynamic nature of healthcare domain, whenever new knowledge is evolved, a new archetype corresponding to a clinical concept needs to be appended to the existing application. Thus, the application handles schema evolution.

For that an auto-detector in our app helps to automatically detect newly added .adl file in repository with the help of "Refresh" functionality. The auto-expander helps to expand the usage of the newly added .adl file by adding them in our app menu under the sub menu "Recent New Files". It has been implemented through the android functionality of "Navigation Drawer". Thus, newly added file can be accessed to generate dynamic interfaces corresponding to each new ADL file. The archetype is then parsed with the help of ADL parser [36] to obtain an Archetype Object Model (AOM) [37] as shown in Fig. 6. AOM defines an archetype in form of an object oriented structure. The AOM is then used by Mapper and GUI Generator. The Mapper and GUI generator consists of set of java classes called wrappers that encapsulate function of the framework. These classes use libraries [39] provided by openEHR on

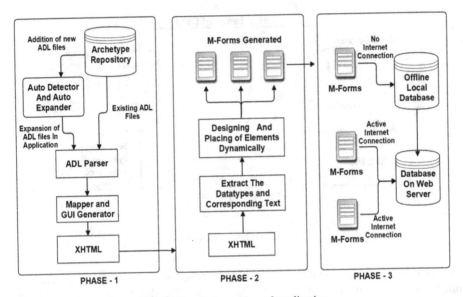

Fig. 5. Detailed working of application

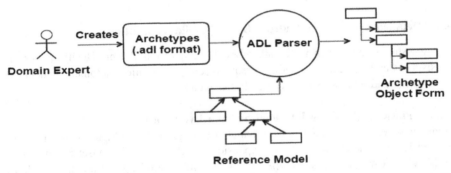

Fig. 6. Conversion of archetype (adl file) to archetype object model

existing Java reference implementation such as adl-parser.jar, adl-serializer.jar and openehr-aom.jar to provide GUI generation functionality. Finally, after mapping and adding GUI components such as text boxes and list boxes corresponding to clinical concepts, .xhtml file is generated as the final output.

4.2.2   Phase 2 (Automatic Generation of M-Forms)

In our study, we aim to provide an interactive user interface for mobile devices. For this purpose, our study considers Android, which is an open source technology and have a very strong java support.

This phase considers .xhtml file generated in first phase as input. The .xhtml file contains useful information about data types required for input fields that have to be

Table 2. Sample interpretation of data types.

Data types in ADL	Data types in XHTML	Interpreted data type
DV_TEXT{*}	inputText value	Edit Text
DV_BOOLEAN	selectBooleanCheckbox	Radio Group with 2 Radio Buttons
C_DV_QUANTITY	InputText id	A number picker widget where we can select a number from a range
DV_TEXT{values {value1,value2..}}	SelectOneMenu	Drop down menu with different choices to select
DV_PROPORTION	inputText value	Edit Text
DV_MUTIMEDIA	link value	Web View for multimedia url

included in M-form. The information is extracted from the .xhtml file and form is generated with the help of interpreted data types as shown in the Table 2.

Table 2 shows sample data types that are used in generation of dynamic interfaces. These data types are most commonly used in archetypes. So, we chose to consider these data types for implementation of proposed application. Asterisk (*) in computer field means "all", So we interpret DV_TEXT (*) as input text value. Thus, any type of data can be entered (for example: 0–9, a–z, A–Z, @, #, $, % etc.). Also Boolean is defined as a variable with two values i.e. 0 and 1. So DV_BOOLEAN is interpreted as data type with two values YES or NO. Similarly C_DV_QUANTITY is being interpreted as a number picker from where a number can be selected within a range. Similarly DV_TEXT {values{value1}} is being interpreted as a menu with limited values to choose from. DV_Proportion is being interpreted as Edit Text in android and is used for precision of decimal point numbers. DV_Multimedia is being interpreted as a web view in android where a URL can be used to render respective multimedia content. Corresponding to new data type in an archetype, a new java class is incorporated in the mobile application. It is with the help of these java classes an archetype after getting parsed from ADL parser generates user interface to fill up the data.

After extracting and interpretation of the data types, the designing and placing of elements takes place that helps to generate m-form. Figure 7 shows generation of M-form from the 'Sample.xhtml' file. With the help of Table 2 one can interpret the data types and attributes of clinical concepts specified in Fig. 7 (in bold and underlined) to corresponding types that is understandable to application as well as application programmer. With the help of M-form user can fill in details which in turn will be stored in the database.

4.2.3   Phase 3 (Generic Persistence of EHRs Database)

EHRs system requires records of patient to be stored in a centralized database to increase the reachability and to enable communication between users. A user can access his/her records or can give access of records to other user around the globe. To enable rapid storing, retrieving and analyzing of records we have used MySQL as a centralized DBMS which is hosted on the web server to provide real time access to records. MySQL is easy to use and understand. It provides support for ACID properties [8] (i.e. Atomicity, Consistency, Isolation and Durability).

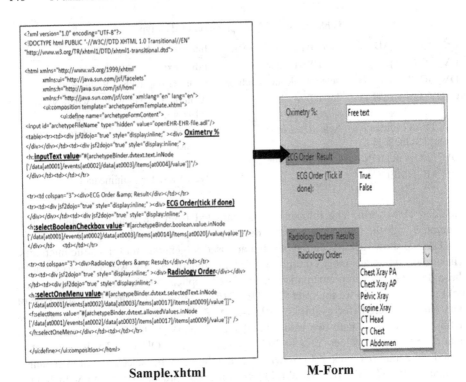

Sample.xhtml   **M-Form**

Fig. 7. XHTML to dynamic M-form

The openEHR specifications aim at accommodating new requirements primarily at the AM (archetype and template) level without the need to change the RM and its associated storage mechanism, thus requiring fewer changes in the application code and persistence mechanism [35]. A generic database schema is required to capture existing and future data requirements. Hence the current research adopts an extended version of Entity Attribute Value (EAV) model. The generic schema of the database is being inspired from EAV model. The approach is referred as node+path persistence.

The process of storing data to centralized data repository requires an active internet connection. However, in absence of internet connectivity, the user response is stored in the local storage of the device within the application. Data present in the local memory of the device can be transferred to the centralized server on regaining internet connectivity.

5   Results

The Sect. 5.1 depicts usability analysis of desktop and mobile application. Section 5.2 shows the screenshots of the Android application. Section 5.3 illustrates different kinds of challenges and their prepared solutions during development of application.

5.1   Usability Analysis of Desktop vs. Mobile Application

Table 3 describes the usability of an application on various parameters, such as data entry, location, storage, and number of users. Some parameters are advantageous in mobile while some have an edge in desktop devices. As depicted in Table 3, in case of mobile devices the most important utility is location. User can access information and fill up the form anywhere with active Internet connection. Unlike existing desktop application (Opereffa), user can add new archetypes with help of refresh functionality that automatically adds it in the navigation drawer/menu of android application. On the other hand, desktop is advantageous to enter data due to big screen size and availability of dedicated keyboard (as hardware). Moreover, desktop provide more storage capacity for storing data offline.

Table 3. Usability analysis of desktop vs. mobile application

Basis	Desktop application	Android application
Data entry	Easier task with a big physical keyboard	Sometimes difficult due to virtual keyboard which is very small in size
Storage	Have high storage capacity. So, large sets of data can be saved locally	Have less storage capacity as compared to desktop. So, most of the data have to be stored on web server
Location	Users have to sit at a particular location to use application on desktop	Users can use mobile while travelling or walking around
Zooming	Required occasionally	Required very frequently
Addition of Archetypes	User has to add archetypes manually and then run the application	User can add archetype at runtime of the application
Users	Single	Multiple
Other Utilities	User can update the records stored in database	Apart from updating, user can delete the records, can book appointments as well as can get reminders

5.2   Screenshots of the Standard Based Android Mobile Application

The screenshot of implementation of phase 1 is shown by Fig. 8. It is showing the navigation drawer or menu having links to already downloaded archetypes, such as clerking, history and examinations. On hitting the refresh button given in the navigation drawer, the application auto detects any newly added archetype in the repository on an online server. If any new archetype is found, application downloads the newly discovered archetype in the device and the layout of the navigation drawer gets changed by adding new adl files under the submenu "Recent NEW FILES" of the navigation drawer.

The screenshot of implementation of phase 2 is shown by Fig. 9. It is showing the dynamic form available to the user after clicking on the archetypes' link from the menu. The data provided by user through forms is stored in a generic database using an active internet connection. The screenshots of implementation of phase 3 are shown by

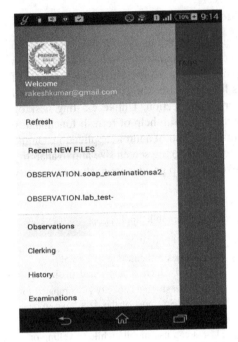

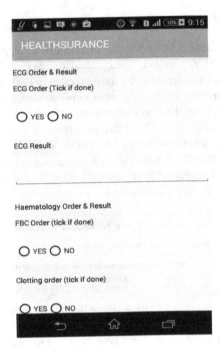

Fig. 8. Navigation drawer with refresh functionality and newly added archetypes and already downloaded archetypes

Fig. 9. Interface of the dynamic form generated

Figs. 10 and 11. With every form submitted, a card view is generated. A card view [11] is a widget in android and it can be displayed on all the platforms of android. The card view in our app shows the name of archetype chosen to create form, status of the form, date and time of submission and a menu comprising of other functions.

A card view corresponding to one form with status "sent" as highlighted and shown in Fig. 10 represents data is successfully stored on an online centralized database server. A card view with status "Not sent" as highlighted and shown in Fig. 11 represents that the data is saved in local storage within application due to no internet connection. "Not Sent" status is updated to "Sent" on tapping 'retry to send' option of the card view and the data gets stored on the online server.

5.3 Challenges and Proposed Solution

The development of the standard based android applications involves many problems that need to be handled to ensure proper working of application. The major problem was the identification of the data types in order to create input fields in the dynamic M-forms. Without the input fields a form is useless. Also the application requires internet to store and retrieve data from database and download necessary files. So, there was a need to have internet access and storage access permissions for the application on android which was done by adding the permissions in manifest file of application.

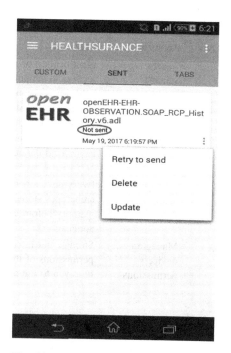

Fig. 10. Generation of Card views when the form is filled and submitted

Fig. 11. Card view showing 'not sent' status along with various options on it

A suitable database was required in order to save the data online. Hence, MySQL database has been chosen on web server provided by free web hosting service. In addition, there has been a need to save the data temporarily if the internet connectivity is lost. Thus, the data is saved in SQLite database of the device's local storage of the healthcare application, which creates a card view with a retry button as well. On hitting retry button, the data will get saved on the online server. Table 4 below summarizes the challenges and the proposed solutions in the current research.

Table 4. Challenges and proposed solutions

S.No	Parameter	Challenge	Proposed solution
1.	Data types differentiation	Identification and corresponding Java definition of different data types in xhtml	Several data types are critically analyzed to generate Java classes to be used for developing interfaces on demand
2.	Database to store and retrieve data	Identification of best suited DBMS	MySQL database is chosen on an online web server

(continued)

Table 4. (*continued*)

S.No	Parameter	Challenge	Proposed solution
3.	Generic Schema	Identification of a generic schema to support dynamic behavior of application and medical knowledge evolution	EAV model has been extended to capture data
4.	Data Persistence	Persisting the data in absence of Internet connectivity to centralized database server	Created a local storage within healthcare application using SQLite database for offline data storage
5.	Storing necessary files	Every device needs to have archetypes stored in its local memory before using the application	The archetypes may be automatically downloaded from the centralized server into the device's local storage
6.	Managing device's permissions	Storage and Internet access permissions of the device must be given to the application	Storage and Internet access permissions has been given to the application programmatically

6   Conclusions

The availability along with the range of mobile devices is increasing at an unbelievable pace. The number of android users worldwide goes around 227 million in 2014 to 352 million in 2016. With its increasing popularity and processing power, the developers are also being attracted to make the experience even more fluent. Although it has small screen size, low processing speed and low memory size as compared to desktops, yet these factors are influential for the developers. This research work aims at the implementation of a mobile application for Standardized Electronic Health Records Database named as Healthsurance. The application make use of archetypes, parse it using ADL parser, generate forms, and thus dynamically generated form saves the data provided by users in a generic database. The data is saved on the centralized server using an active internet connection, while precaution has been taken in case the internet connectivity is not there.

In this paper, the authors have been inspired by the previous works done on EHRs, archetypes, the desktop version of the clinical application (Opereffa). Consequently, an approach has been given for the adl parsing in the android devices with low processing speed, low memory size with saving of data on the centralized server. The authors have successfully implemented the application on a sample set of archetypes and shown the working of the application with the help of various phases as described in Sect. 4.

The type of users who will be benefitted from the android app includes doctors and patients. They will be benefitted on the basis of interoperability, usability and persistence.

- **Interoperability** - The earlier scenario was that the information sharing among hospitals was not possible due to storing of data in different formats. But due to the introduction of EHRs based on the standards of semantic interoperability, it is possible that patients can share their history, examinations' report to other hospitals as well.
- **Persistence of Data** - To capture data, a generic database termed as Entity Attribute Value (EAV) is adopted. Generic nature of database enables storing any data without prior information of the participating attributes. Thus, it supports dynamic behavior of proposed application.
- **Usability** - Any update in medical knowledge is reflected in terms of versioning of archetypes without making any changes in the application code and database schema. Also the app has nice transitions like navigation drawer for hassle free navigation. As the data of user is very precious, it should not be lost. Thus, the application has the feature of storing the data offline when no internet connection is there. Moreover, the refresh functionality helps to auto detect new archetypes (adl files) in archetype repository and download them in the application, hence expanding the usage of .adl files.

Currently, application is built for the android devices. In future, authors aim to provide support for other operating systems, such as Symbian, windows and iOS as well.

References

1. Electronic health record (2017). https://en.wikipedia.org/wiki/Electronic_health_record. Accessed 15 Jan 2017
2. Parashar, H.J., Sachdeva, S., Batra, S.: Enhancing access to standardized clinical application for mobile interfaces. In: Madaan, A., Kikuchi, S., Bhalla, S. (eds.) DNIS 2013. LNCS, vol. 7813, pp. 212–229. Springer, Heidelberg (2013). doi:10.1007/978-3-642-37134-9_17
3. Abel, D., Gavidi, B., Rollings, N., Chandra, R.: Development of an Android Application for an Electronic Medical Record System in an Outpatient Environment for Healthcare in Fiji: arXiv:1503.00810v1 (2015)
4. ISO 13606. https://www.iso.org/standard/50122.html. Accessed 15 Jan 2017
5. Ventola, C.L.: Mobile devices and apps for health care professionals: uses and benefits. Pharm. Ther. **39**(5), 356 (2014)
6. Martínez-Costa, C., Menárguez-Tortosa, M., Fernández-Breis, J.T.: An approach for the semantic interoperability of ISO EN 13606 and OpenEHR archetypes. J. Biomed. Inform. **43**(5), 736–746 (2010)
7. Archetype Definition Language ADL 2. http://www.openehr.org/releases/trunk/architecture/am/adl2. Accessed 15 Jan 2017
8. ACID. https://en.m.wikipedia.org/wiki/ACID. Accessed 15 Jan 2017
9. Entity–attribute–value model. https://en.m.wikipedia.org/wiki/Entity–attribute–value_model. Accessed 15 Jan 2017
10. JavaServer Faces. https://en.wikipedia.org/wiki/JavaServer_Faces. Accessed 15 Jan 2017
11. Material Design for Android. https://developer.android.com/design/material/index.html. Accessed 15 Jan 2017
12. PostgreSQL. https://www.postgresql.org/about/. Accessed 15 Jan 2017

13. Global smartphone sales to end users from 1st quarter 2009 to 3rd quarter 2016, by operating system (in million units). https://www.statista.com/statistics/266219/global-smartphone-sales-since-1st-quarter-2009-by-operating-system/. Accessed 15 Jan 2017

14. CKM. Clinical Knowledge Manager. http://www.openehr.org/ckm/. Accessed 15 Jan 2017

15. Number of mobile phone users worldwide from 2013 to 2019 (in billions). https://www.statista.com/statistics/274774/forecast-of-mobile-phone-users-worldwide/. Accessed 15 Jan 2017

16. Electronic Health Records Software User Report – 2014. http://www.softwareadvice.com/resources/ehr-software-user-trends-2014/. Accessed 15 Jan 2017

17. Sutcliffe, A., Ryan, M., Doubleday, A., Springett, M.: Model mismatch analysis: towards a deeper explanation of users' usability problems. Behav. Inform. Technol. **19**(1), 43–55 (2000)

18. Yuan, W.: End-user searching behavior in information retrieval: a longitudinal study. J. Assoc. Inform. Sci. Technol. **48**(3), 218–234 (1997)

19. Jagadish, H.V., Chapman, A., Elkiss, A., Jayapandian, M., Li, Y., Nandi, A., Yu, C.: Making database systems usable. In: Proceedings of the 2007 ACM SIGMOD International Conference on Management of Data, pp. 13–24, June 2007

20. Bharaj, T.S., Sachdeva, S., Bhalla, S.: AsthmaCheck: multi-level modeling based health information system. In: Wang, F., Yao, L., Luo, G. (eds.) DMAH 2016. LNCS, vol. 10186, pp. 139–154. Springer, Cham (2017). doi:10.1007/978-3-319-57741-8_9

21. openEHR. http://www.openehr.org/. Accessed 15 Jan 2017

22. Ocean Informatics. https://code4health.org/_attachment/modellingintro/2015_11_Modelling_Intro.pdf. Accessed 15 Jan 2017

23. Poll Results- Top 10 archetypes for use in an Emergency – Health Information Model – opeEHR wiki. https://openehr.atlassian.net/wiki/display/healthmod/Poll+Results+-+Top+10+archetypes+for+use+in+an+Emergency. Accessed 15 Jan 2017

24. Patrick, J., Ly, R., Truran, D.: Evaluation of a persistent store for openEHR. In: Proceedings of the HIC 2006 and HINZ 2006, p. 83 (2006)

25. Nadkarni, P.M., Brandt, C., Frawley, S., Sayward, F.G., Einbinder, R., Zelterman, D., Schacter, L., Miller, P.L.: Managing attribute–value clinical trials data using the ACT/DB client–server database system. J. Am. Med. Inform. Assoc. **5**(2), 139–151 (1998)

26. Brandt, C.A., Nadkarni, P., Marenco, L., Karras, B.T., Lu, C., Schacter, L., Fisk, J.M., Miller, P.L.: Reengineering a database for clinical trials management: lessons for system architects. Control. Clin. Trials **21**(5), 440–461 (2000)

27. Nadkarni, P.M., Marenco, L., Chen, R., Skoufos, E., Shepherd, G., Miller, P.: Organization of heterogeneous scientific data using the EAV/CR representation. J. Am. Med. Inform. Assoc. **6**(6), 478–493 (1999)

28. Shepherd, G.M., Healy, M.D., Singer, M.S., Peterson, B.E., Mirsky, J.S., Wright, L., Smith, J.E., Nadkarni, P., Miller, P.L.: SenseLab: a project in multidisciplinary. Neuroinform. Overview Hum. Brain Proj. **1**, 21 (1997)

29. Duftschmid, G., Wrba, T., Rinner, C.: Extraction of standardized archetyped data from Electronic Health Record systems based on the Entity-Attribute-Value Model. Int. J. Med. Inform. **79**(8), 585–597 (2010)

30. Johnson, S.B.: Generic data modeling for clinical repositories. J. Am. Med. Inform. Assoc. **3**(5), 328–339 (1996)

31. Muñoz, A., Somolinos, R., Pascual, M., Fragua, J.A., González, M.A., Monteagudo, J.L., Salvador, C.H.: Proof-of-concept design and development of an EN13606-based electronic health care record service. J. Am. Med. Inform. Assoc. **14**(1), 118–129 (2007)

32. Frade, S., Freire, S. M., Sundvall, E., Patriarca-Almeida, J.H., Cruz-Correia, R.: Survey of openEHR storage implementations. In: 2013 IEEE 26th International Symposium on Computer-Based Medical Systems (CBMS), pp. 303–307, June 2013

33. Wang, L., Min, L., Wang, R., Lu, X., Duan, H.: Archetype relational mapping-a practical openEHR persistence solution. BMC Med. Inform. Decis. Making 15(1), 88 (2015)

34. Node+Path persistence. https://openehr.atlassian.net/wiki/pages/viewpage.action?pageId= 6553626. Accessed 15 Jan 2017

35. Atalag, K., Yang, H.Y., Tempero, E., Warren, J.R.: Evaluation of software maintainability with openEHR–a comparison of architectures. Int. J. Med. Inform. 83(11), 849–859 (2014)

36. Java, ADL Parser Guide – projects - openEHR wiki. http://www.openehr.org/wiki/display/ projects/Java+ADL+Parser+Guide. Accessed 15 Jan 2017

37. Beale, T.: The openEHR archetype model: archetype object model. In: The Openehr Release 1.0.2, openEHR Foundation (2008)

38. Dinu, V., Nadkarni, P.: Guidelines for the effective use of entity–attribute–value modeling for biomedical databases. Int. J. Med. Inform. 76(11), 769–779 (2007)

39. openEHR Java Libraries. https://github.com/openEHR/java-libs?files=1. Accessed 15 Jan 2017

40. Opereffa Project – Projects – openEHR wiki, http://www.openehr.org/wiki/display/projects/ Opereffa+Project. Accessed 15 Jan 2017

Author Index

Akbilgic, Oguz   73
Araki, Kenji   88

Batra, Shivani   136
Begoli, Edmon   41
Bernal-Rusiel, Jorge   29
Bhalla, Subhash   136
Bhargava, Sagar   136

Caetano, Bernardo   102
Chen, Yiru   9
Christian, J. Blair   41

Dubovitskaya, Alevtina   3

Farinha, Rui   102
Fonseca, Manuel J.   102

Gadepally, Vijay   41
Galhardas, Helena   9, 102
Golab, Lukasz   55
Grant, P. Ellen   29

Haehn, Daniel   29
Honda, Yuichi   88

Jain, Naman   136
Jain, Prateek   136

Kamaleswaran, Rishikesan   73
Kathiravelu, Pradeeban   9
Krieger, Orran   29
Kushima, Muneo   88

Lages, Nuno F.   102

Mahajan, Ruhi   73

Papadopoulos, Stavros   41
Pereira, João D.   102
Pienaar, Rudolph   29

Qin, Zhaohui   36

Rannou, Nicolas   29
Ryu, Samuel   3

Sachdeva, Shelly   136
Schumacher, Michael   3
Sharma, Ashish   9
Sun, Xiaobo   36

Toulis, Andrew   55
Turk, Ata   29

Van Roy, Peter   9
Veiga, Luís   9

Wang, Fusheng   3, 36, 121
Wang, Yu   121

Xu, Zhigang   3

Yamazaki, Tomoyoshi   88
Yokota, Haruo   88

Author Index

Printed in the United States
By Bookmasters